EARTHQUAKES

Trapped victims being rescued from collapsed upper-level of I-880 Cypress Street two-tier overpass, October 1989 Loma Prieta Earthquake.
Source: Roy H. Williams, *Oakland Tribune.*

EARTHQUAKES

An Architect's Guide to Nonstructural Seismic Hazards

HENRY J. LAGORIO
Professor of Architecture
University of California, Berkeley

Site Planning
Building Design
Nonstructural Components
Urban Planning and Design
Rehabilitation of Existing Buildings
Disaster Recovery and Reconstruction

A Wiley-Interscience Publication
JOHN WILEY & SONS, INC.
New York / Chichester / Brisbane / Toronto / Singapore

In recognition of the importance of preserving what has been written, it is a policy of John Wiley & Sons, Inc. to have books of enduring value published in the United States printed on acid-free paper, and we exert our best efforts to that end.

Library of Congress Cataloging-in-Publication Data:

Lagorio, Henry J.
Earthquakes: An Architect's Guide to Nonstructural Seismic Hazards / Henry J. Lagorio.
p. cm.
"A Wiley-Interscience publication."
Includes bibliographical references.
1. Earthquake resistant design. I. Title.
TA658.44.L34 1990
624.1'762—dc20 90-35605
ISBN 0-471-63302-X CIP

Printed in the United States of America

10 9 8 7 6 5 4 3 2 1

To three "earthquake greats":
S. T. Algermissen, the late H. J. Degenkolb, and K. V. Steinbrugge,

and

To my wife, Natalie,
who was always at my side with encouragement, support, and
ready to "shake things up."

CONTENTS

PREFACE

Primarily, this book is written for members of the architectural profession as a means of transferring to them some of the latest developments in earthquake hazards reduction. It is currently recognized that earthquakes are a national problem that all design professionals must deal with in meeting the challenges created by ever increasing urban populations.

My interest in earthquake engineering began in 1947 when I joined the faculty of architecture at the University of California at Berkeley, and became fascinated by the architectural implications of seismic safety. This interest was considerably strengthened by two other professors who joined the architecture department in the 1950s, the late George Simonds, an architect, and Karl V. Steinbrugge, a structural engineer. As these two had considerably more experience than I in earthquake engineering, their strong commitment to seismic safety reinforced and confirmed our common interests on a multidisciplinary basis. Both are acknowledged for their pioneering contributions to architectural planning and design from a seismic overview.

In the early 1970s, my sensitivity to seismic safety concerns was further heightened when I was selected to join a multidisciplinary team that developed methodology for the very first earthquake vulnerability analysis study of a major metropolitan area. The final report, "A Study of Earthquake Losses in the San Francisco Bay Area: Data and Analysis," was issued in 1972 under the auspices of the National Oceanic and Atmospheric Administration (NOAA), (see Algermissen S. T., et. al, 1972). Members of that team are herein acknowledged for their creative efforts in bringing the innovative study to realization: S. T. Algermissen, W. A. Rinehart, James Dewey, Karl V. Steinbrugge, Henry J. Degenkolb, Lloyd S. Cluff, Frank E. McClure, Stanley Scott, Robert A. Olson, and Richard F. Gordon.

Acknowledgment is also given to the following individuals who have made important contributions to earthquake hazards mitigation goals in the interests of public health and safety: Mihran S. Agbabian, Richard Andrews, Clarence R. Allen, William A. Anderson, Alfredo H. S. Ang, Christopher Arnold, Samuel Aroni, J. Marx

Ayers, James E. Beavers, Vitelmo V. Bertero, Marta Blair-Tyler, John A. Blume, Bruce A. Bolt, Roger D. Borcherdt, Elmer E. Botsai, Jane Bullock, Vincent R. Bush, Mehmet Celebi, Anil K. Chopra, Ray W. Clough, Mary C. Comerio, Louise K. Comfort, LeRoy L. Crandall, James F. Davis, Neville C. Donavan, Russell Dynes, Richard K. Eisner, Eric Elsesser, Luis E. Escalante, Paul Flores, Nicholas F. Forell, James H. Gates, Peter Gergely, Melvyn Green, Marjorie R. Greene, Teresa Guevara, William J. Hall, Robert D. Hanson, Walter W. Hays, David B. Helfant, George W. Housner, I. M. Idriss, Jesus Iglesias, Paul C. Jennings, Donald K. Jephcott, James D. Jirsa, Barclay G. Jones, Carl B. Johnson, Edwin H. Johnson, Gary D. Johnson, Roy G. Johnston, Earle W. Kennett, Helmut Krawinkler, Frederick Krimgold, Julio Kuroiwa, Joe J. Litehiser, Shi Chi Lui, George G. Mader, Stephen A. Mahin, John Meehan, Roberto Meli, David L. Messinger, Ugo Morelli, Paul Neel, Joseph P. Nicoletti, Joanne M. Nigg, Eric K. Noji, Gordon B. Oakeshott, Robert A. Olson, Richard A. Parmelee, Joseph Penzien, Chris D. Poland, Igor P. Popov, Robert F. Preece, Jane Preuss, Robert K. Reitherman, Christopher Rojahn, Badaoui M. Rouhban, Roger E. Scholl, Haresh C. Shah, John B. Scalzi, Anshel J. Schiff, Earl Schwartz, Raymond B. Seed, Roland L. Sharpe, Robin Shepherd, Carl J. Stepp, James L. Stratta, Charles C. Thiel, Thomas L. Tobin, Stephen Tobriner, Tousson R. Toppozada, Susan Tubbesing, Robert A. Uhrhammer, Anestis S. Veletsos, Robert E. Wallace, M. Wang, Robert V. Whitman, Kit M. Wong, Thomas D. Wosser, Richard N. Wright, Loring A. Wyllie, and Theodore Zsutty.

I also acknowledge the contributions made by many others too numerous to mention here due to space limitations, who freely shared their knowledge, photographs, and published materials with me over the years. Every effort has been made to identify data received to its source, including credits for drawings and photographs. Undoubtedly I have not properly credited some for photographs, technical papers, or other materials received. I offer my sincere apologies to those whom I have missed.

Finally, I wish to thank Nora Watanabe and Kellie Crockett of the Center for Environmental Design Research (CEDR) at the University of California, Berkeley, for their generous assistance in computer formatting techniques and welcomed clerical support.

HENRY J. LAGORIO

Berkeley, California
September 1990

EARTHQUAKES

INTRODUCTION

A severe, damaging earthquake is one of the most terrifying and devastating events that can be experienced. The energy released by a violent seismic event, such as the 1964 Great Alaska earthquake, with an 8.6 magnitude on the Richter scale, is said to be equal to a nuclear explosion of over 1000 megatons.

Ever since the celebrated 1906 San Francisco earthquake, which resulted in 700 deaths followed by an urban conflagration that destroyed 521 blocks of the downtown area in three days, and the 1923 Kanto (Tokyo), Japan, earthquake, compounded by a devastating fire in which 143,000 lives were lost, the destructive power of an earthquake followed by powerful secondary effects has held the public spellbound and inspired scientists to new research efforts. Fascinating reports of historic earthquakes appeared as early as 63 A.D. in Naples, Italy, culminating with accounts of the great eruption of Mt. Vesuvius that buried Pompeii in 79 A.D. Pliny the Younger aptly recorded the effects of the disaster in his letters to the historian Tacitus. In 1936, even Hollywood had reacted to our interest in earthquakes by producing several motion pictures on the 1906 San Francisco event, the most popular being *San Francisco* starring Clark Gable. Dramatically, 83 years later, on October 17, 1989, at 5:04 P.M., the entire San Francisco Bay Area region was shaken by another major earthquake, measuring a 7.1 magnitude on the M_s scale, which caused severe damage in seven counties around the bay. The following day, when San Francisco's mayor, Art Agnos, was told by reporters that leading seismologists in the area did not consider it to be **"THE BIG ONE"** in comparison to the 1906 San Francisco earthquake, which had an 8.3 magnitude, he could only reply, "I don't know about that, but this one was **BIG ENOUGH** for me!!!" For detailed information on the October 1989 Loma Prieta earthquake in the San Francisco Bay Area region, refer to Chapter 12.

In more recent years, a popular adage among design professionals was "Earthquakes don't kill people, buildings do." Yet, despite this sentiment, in the United States the first specific seismic provisions for the earthquake-resistant design of buildings did not appear in the building code until 1934 in California, a year after the 1933 Long Beach earthquake which resulted in property losses of $41 million and 120 deaths. It is sobering to realize that many pre-1934 unreinforced masonry buildings, in their original design state, never having undergone any seismic upgrading or rehabilitation, are still in use today in areas of high seismic risk around the country. California alone is estimated to have over 50,000 such structures throughout various municipalities.

Until the early 1950s, principal efforts in earthquake-preparedness planning, hazards investigations, and research relative to building performance standards were considered the realm of earth scientists, seismologists, geophysicists, and civil/structural engineers. The goals, in order, were simply to avoid casualties, preclude catastrophic collapses of buildings, minimize severe structural damage, and reduce property losses. Comparatively speaking, less attention and emphasis were placed on the application of earthquake engineering research to the architectural design aspects of the problem.

However, in the early 1950s several architects started informal investigations on their own regarding the architectural aspects of the seismic performance of buildings. Yet, it was not until 1964, after the Great Alaska earthquake, which destroyed many buildings in Anchorage, Seward, and Valdez, that the importance of architectural considerations in earthquake hazards mitigation programs was clearly delineated. Two of the first scientifically based investigations completed at the time were by an architectural consultant, Harold D. Hauf, and a mechanical engineer, J. Marx Ayres, who pinpointed specific seismic problems related to the architectural aspects of building design and the performance of nonstructural building elements. It was discovered that damage to the architectural, nonstructural elements of a building system during an earthquake could account for up to 65 to 70 percent (upper limits) of a building's replacement costs. At this level of damage, even though the basic structural system may have remained relatively intact, repair costs would have made rehabilitation of the building highly questionable if not downright unfeasible. As another measure of architectural considerations studied at the time, the proportion of property loss traceable to the damage of a building's contents was estimated on average to reach as high as 35 percent of the total property losses, and that was in 1964. As high as that may seem, the reader should remember that California's Silicon Valley, where high-tech buildings designed for the manufacture and storage of very expensive electronic data-processing equipment are located, did not exist at the time. In today's computer-oriented environment, the loss of such a building's contents could easily cost more than twice that of the structure itself.

By the time of the 1971 San Fernando earthquake, it had become quite clear that damage to the architectural, nonstructural elements of a building and its contents could result in a serious number of casualties, total building impairment, and major economic losses, even though the basic structural system may not have been significantly damaged.

Today, earthquake hazards mitigation procedures represent a research area that clearly is part of the architect's responsibility for public health and safety. For an architect to be uninvolved in the study of this critical topic area is an unacceptable situation relative to professional liability concerns. In addition, in this day and age of international travel by jet airplanes, and supersonic ones at that, sooner or later, no matter where the architect's home office is located, a client will appear with a project to be designed for a site located in a seismically active area in the United States or abroad. And, as the team leader, it is the architect who faces the greatest exposure to professional liability, even when the engineering consultants are given total responsibility for the earthquake-resistant structural design.

Developed as a "primer" by the American Institute of Architects in 1975, *Architects and Earthquakes* (Botsai et al., 1976) was the first such publication ever to focus specifically on the overall responsibilities that the architectural profession has in the study of earthquakes and their effects on a building's performance. There has been further research and new architectural lessons and approaches to the seismic safety problem since this author participated in that first publication on the architectural concerns of earthquake-resistant design.

Later, the first detailed accounting of the critical influence that the architect's preliminary design of a building's configuration has on its seismic performance was published by John Wiley & Sons through an architect, Christopher Arnold, AIA (see Arnold and Reitherman, 1982); it was the result of research grant sponsored by the National Science Foundation in Washington, D.C., in 1977 under the guidance of Dr. John Scalzi in 1977, a program director there. In the 1980s, for the first time, there was full-scale laboratory testing of the seismic performance of exterior cladding materials and interior partitions in accommodating earthquake-induced deformations in a steel moment frame under interstory drift limitations. The testing, which took place in Tsukuba, Japan, was part of the U.S./Japan Cooperative Research Program with funding also supported by an NSF grant to a U.S. architect/engineer, Marcy Wang (1987).

As these episodes take their place in the historic record, one objective of this book is to bring to the architectural community these and other new developments in earthquake hazards mitigation procedures that have a decisive impact on earthquake-resistant building construction and design. Figure 1 shows a seismic map indicating the location and distribution of historic earthquakes in the United States. Accordingly, another purpose of this book, in its architectural orientation, is to cover some of the latest information

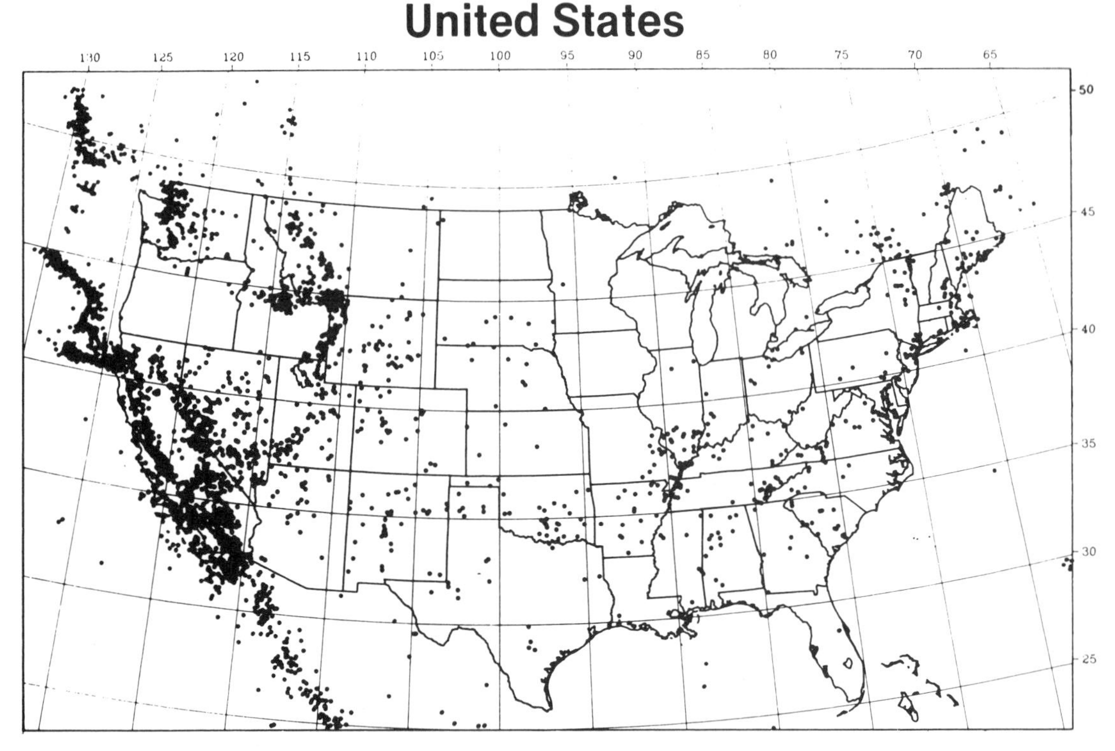

Figure 1 Location of all historic earthquakes of magnitude 5.5 or larger; all earthquakes of magnitude 5 to 5.4 since 1925; all recorded earthquakes of magnitude 4 to 4.9 since 1962; and all recorded earthquakes of magnitude 3.5 to 3.9 since 1975 in the United States. *Source:* U.S. Geologic Survey.

garnered, since this map was made, from the numerous investigations of recent earthquakes, many of which were personally examined by the author in field studies throughout the world. It is hoped that by the time that architects finish reading this publication, they will have a comprehensive and basic understanding on how and in what way earthquakes impact their professional practice.

CHAPTER

1 EARTHQUAKE CAUSES AND EFFECTS

Typically, when the subject of seismic safety is discussed, the first questions often asked by clients are, "What are the causes of earthquakes?" "How often do they occur?" and "What will they do to my building?" Answers to these queries are not easy, since new scientific data, which modify previous opinions, become available after each major damaging earthquake. Among them has been the further development, during the last two decades, of plate tectonic theory, which has led to a greater understanding of earthquake causes, effects, occurrences, and impacts on the built environment.

Centuries before, and until the science of seismology became formalized, the supernatural, in one form or another, was often cited as the cause of seismic events. Even today, many in Japan believe earthquakes are due to the wild thrashing of the great Giant Catfish (Namazu), who dwells deep within the earth, trying to reach the surface. Japanese pictorial representations and woodblock prints of the eighteenth and nineteenth centuries (Figure 10-4) depict the Kashima deity pinning down the Giant Catfish with a heavy stone or heavy steel shaft so that it will not succeed in rising to the surface and causing an earthquake.

Other mythological earthquake legends still exist within many cultures. In Assam, the Kukis believe that another race lives inside the earth and shakes the ground whenever they become curious enough to see if anyone is still living on top. Whenever the Kukis feel the ground shaking, they immediately shout, "Alive, alive!!" to ensure the race below that someone is definitely alive on the surface above. The Kaffirs in Mozambique envision the earth catching a severe fever followed by recurring chills and fits of momentous shivering.

Early Greek philosophers, including Aristotle, were among the first to hypothesize that there must be a natural, rational, physical reason for earthquakes. They seriously theorized about the causes of earthquakes, but their explanations were very crude and quite often even further from the truth than the primitive legends. Aristotle concluded that earthquakes were generated by fierce pockets of expansive gases imprisoned in subterranean cavities; these gases, in their violent struggle to escape through ruptured fissures, supposedly viciously vibrated the earth.

In 1668, Robert Hooke published his *Discoveries on Earthquakes,* in which he theorized that earthquakes were caused by the shrinking of the earth's surface. After the 1755 Lisbon earthquake, priests in Portugal were sent instructions to send in their accounts of the event by indicating the exact time and direction of ground motions and a tally of the damage in their areas. In effect, this became one of the first earthquake reconnaissance investigations to be conducted on a statistical basis.

In the eighteenth century, Frezier cited his assumptions on the causes of earthquakes in his records of voyages along the coast of Chile and Peru during a three-year period, 1712–1714. He discusses relationships between salt, sulfur, and metal mines in the region, including "burning mountains" (volcanoes), rainfall, and other potential triggers of earthquakes. In 1862, Robert Mallet's account of the 1857 Neapolitan earthquake was one of the first detailed attempts to analyze the effects of earthquakes from a scientific perspective. Although his analytical reports of the geological aspects of earthquakes and their effects on buildings were somewhat simplified, they nonetheless represented an advancement in the study of seismic events. He also recommended that recording stations, or "observatories," be established throughout the world as a means of collecting data on the earthquake phenomenon.

In 1880, the Seismological Society of Japan was established. After the great earthquake of Nohbi in 1891, an Imperial Earthquake Investigation Committee was formed to study the effects of that earthquake and the potential impact of future seismic events. In addition, recommendations were made to establish earthquake-resistant design methods for structures and facilities.

The 1906 San Francisco earthquake was an event that received detailed study by many scientists, including engineers, geophysicists, and seismologists. The San Andreas Fault was carefully investigated, as were the mechanisms of earthquakes. The elastic rebound theory of faulting, which remains one of the basic explanations of stress buildup along a fault, was derived from these studies. (The elastic rebound theory is covered in detail later in this chapter.)

GENERAL THEORY OF EARTH MOVEMENTS: PLATE TECTONICS

The theory of plate tectonics, introduced in 1967, asserts that the mantle, or upper crust, of the earth is in constant motion as segments of its lithosphere, technically referred to as "plates," slowly, continuously, and individually slide over the earth's interior. Originally, the crust of the earth was held to be a single mass, one huge supercontinent without the existence of any ocean basins. About 200 million years ago, this supercontinent started to gradually split apart and drift into the segments (plates) of land masses and oceans with which we are familiar today. A total of six major plates and over six smaller minor plates are said to make up this system of the upper mantle. These plates meet in "convergent zones" and are pushed apart in "divergent zones." (See Figure 1-1.)

At zones of divergence, molten rock from beneath the crust gushes up to fill in the resulting rift and forms a ridge. As these ridges are formed, pressures from their buildup are said to be responsible for the creation of spreading plate boundaries shown in Figure 1-1. This phenomenon has been observed as occurring at midocean locations exemplified by the Mid-Atlantic Ridge and the Mid-Pacific Rise. The Red Sea is often cited as an example of a

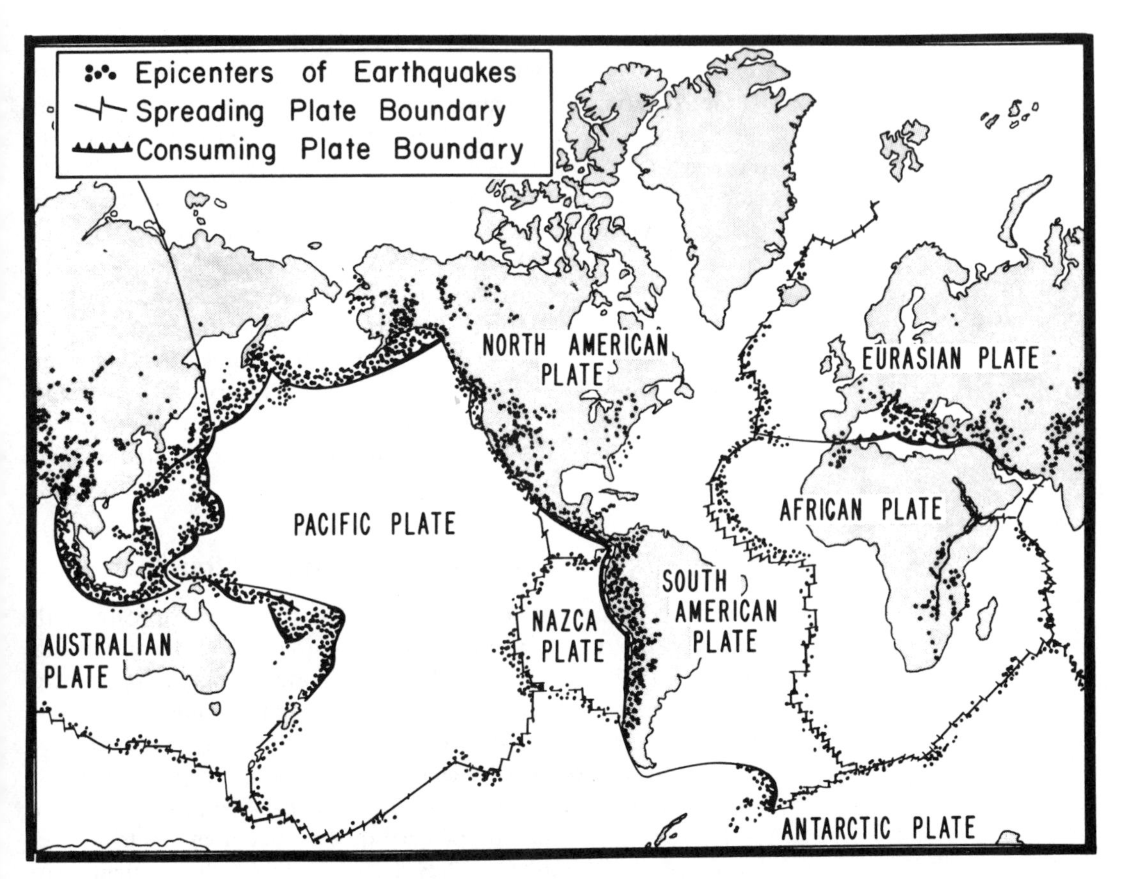

Figure 1-1 World map of seismic zones and major crustal plates. *Source:* U.S. Geological Survey.

young spreading ridge, as it separates Africa from the Arabian Peninsula.

At zones of convergence, subduction occurs as one plate slides under the other, forming a trench as it returns material from the leading edge of the lower plate to the earth's interior. The Aleutian Trench is an example of a subduction zone. Plates can also slide past each other laterally, as well as rotate, as they are pressed against each other as one or both plates may move relative to one another. This phenomenon may be illustrated by the Pacific Plate, bordering on the West Coast of the United States, as it moves northeasterly past the North American Plate along the zone of the San Andreas Fault in California. This movement along the San Andreas Fault has been measured, with laser beams on either side of the fault system, to be at an average rate of 1.5 to 2.5 inches per year.

Recently, however, in the late 1980s, the geotechnical portrayals of the earth's surface have changed. Contrary to earlier general views that the earth's crust was composed of thin fragmented plates that floated on the mantle below, some research geophysicists at Cambridge University now believe that the continents may be much thicker, with sections of dense rock formations, like jagged teeth in a jawbone, which are deeply anchored to the lower mantle. Their research is indicating that the earth's outer core contains mountains and valleys as high and deep as those that we see on the surface. One of the goals of the scientists studying the earth's interior is to solve the last major puzzle in plate tectonics: Why do the plates move as they do? What caused the earth's crust to break apart into pieces, and what pushes the pieces around the earth's surface? Most geologists theorized that the "much-slower-than-molasses" churning motions of molten rock in the earth's mantle are the "forces that drive the motion of the plates," but according to John Woodhouse (1989), a geophysicist at Harvard College who has worked on an imaging project of the earth's interior, "to understand exactly why and how this takes place, more detailed pictures of the mantle are needed."

Earthquakes at Plate Boundaries and Within Plates

As indicated, it is implied that these plate movements create earthquakes as the respective plates collide with one another and/or subduct one under another in subduction zones. Ninety percent of all earthquakes occur in the vicinity of these plate boundaries. Where plates push into each other and one slides beneath the other, shallow to deep-seated earthquakes are common. Deep-seated earthquakes are less likely to occur where plates slide past each other.

The other 10 percent of earthquakes occur at faults located within plates (intraplate). They are much less frequent than those at plate boundaries, and the seismic mechanism is not as well understood. Earthquakes of this genre are found in the midwest and the eastern regions of the United States, the most notable being a series of major seismic events during 1811–1812 in the New Madrid region of Missouri and the Mississippi River basin. A recent theory proposes that these earthquakes occur along the hidden, deep-seated seams of large land masses that had been pushed together ages ago during the early formations of the continents as they are known today. In any event, such severe internal seismic activities disclose the fact that these tectonic intraplate zones are not totally free of internal rupture.

Plate movements in subduction zones such as those found along the western coasts of Washington State and Mexico have been cited as causing major earthquakes. Such an event involving the Cocos Plate was responsible for the 1985 Mexico/Michoacán earthquake, which was accountable for the major damage in Mexico City. (See Figure 1-2.)

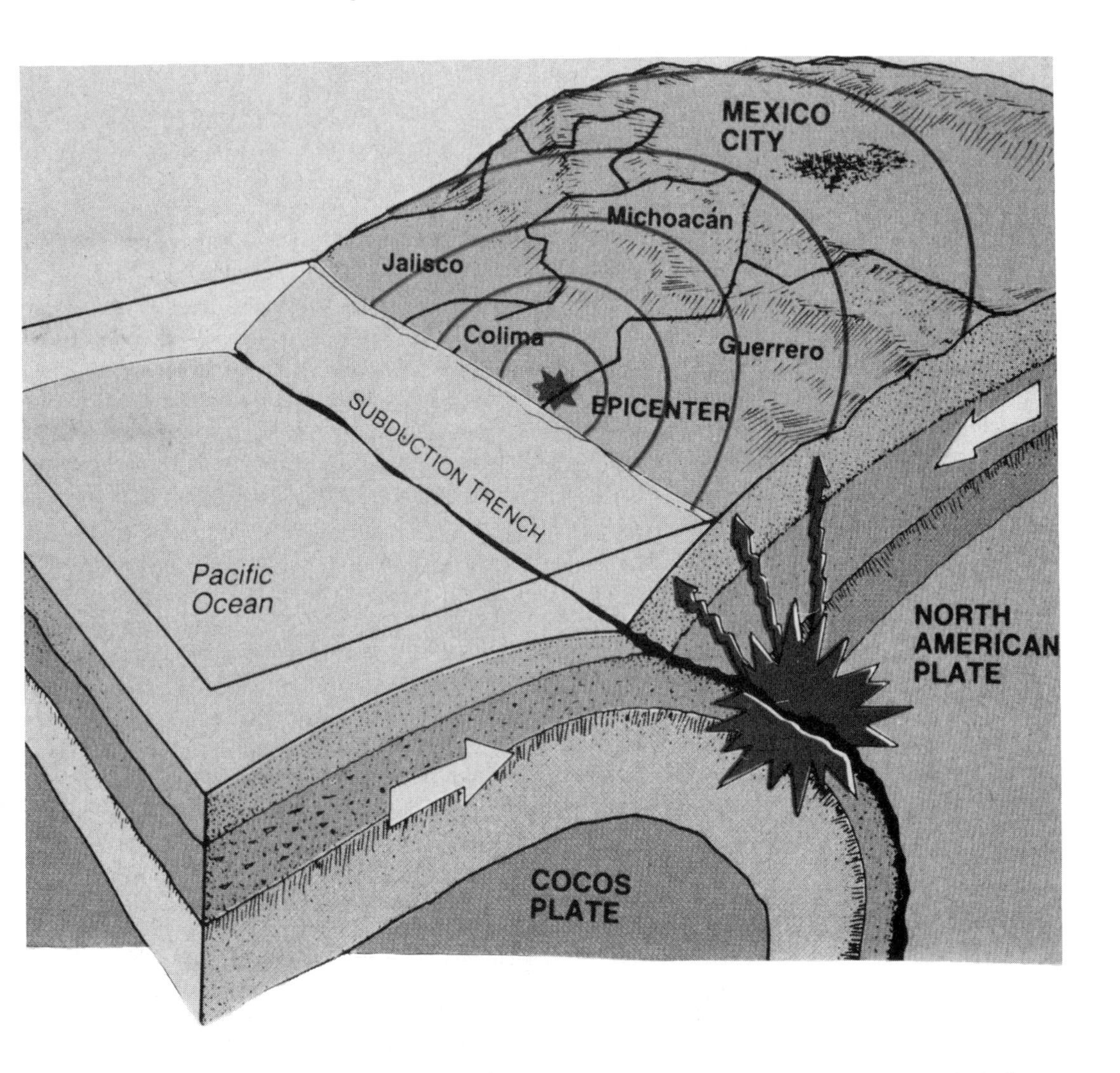

Figure 1-2 Cocos Plate subduction zone at west coast of Mexico.

THE BASIS OF EARTHQUAKES

As part of the earth's upper crust move tectonically relative to one another in an extremely slow exchange, the result may be either a sluggish creeping movement or a gradual accumulation of stress along a fractured fault zone. Creep movements as dèscribed above have been clearly observed in many areas and may be measured by the gradually increasing offsets seen in curbs, fences, streams, and even in buildings when constructed in locations straddling an active fault. One example of this manifestation is readily seen across the Hayward Fault Zone on the eastern side of the San Francisco Bay Area. When a portion of the fault is "locked into place" so that the creep is not allowed to occur along that portion of the fault zone, it is believed that an accumulation of stress builds up until it exceeds the strength holding the locked portion in place, causing a rupture that produces the intense vibrations associated with an earthquake.

Types of Earthquake Faults

There are various types of earthquakes faults around the world. Displacement along these fault zones during an earthquake can be quite dramatic and may result in diverse offsets: vertical, horizontal, or a combination of the two. Along the San Andreas Fault, which preliminary analysis reveals is the culprit in the October 1989 San Francisco Bay Area regional earthquake, movement is characteristically horizontal in a strike–slip motion. In contrast, however, large vertical offsets occurred as a result of fault action during the 1971 San Fernando earthquake. Figure 1-3 indicates the various types of earthquake faults, simply classified by the configuration and direction of offset, or slip, as shown by the respective drawings.

Elastic Rebound Theory

Following the 1906 San Francisco earthquake, Harold Fielding Reid proposed the elastic rebound theory, Steinbrugge (1982). Based on geological evidence, including triangulation surveys by the U.S. Coast and Geodesic Survey across the region traversed by the 1906 fault break and his own laboratory experiments, Reid submitted that

> it is impossible for rock to rupture without first being subjected to elastic strains greater than it can endure. We conclude that the crust in many parts of the earth is being slowly displaced, and the difference between displacements in neighboring regions sets up elastic strains, which may become stronger than the rock can endure. A rupture then takes place and the strained rock rebounds under its own elastic stresses, until the strain is largely or wholly relieved. In the majority of cases, the elastic rebounds on opposite sides of the fault are in opposite directions.

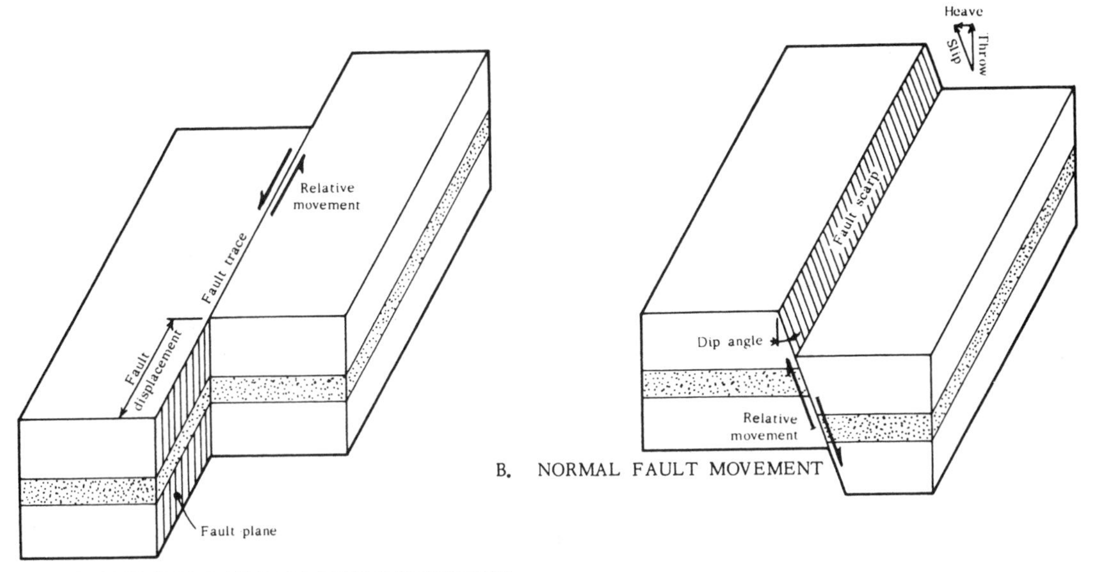

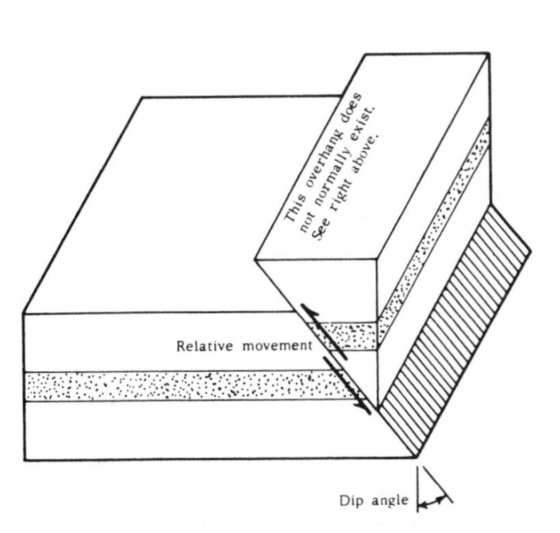

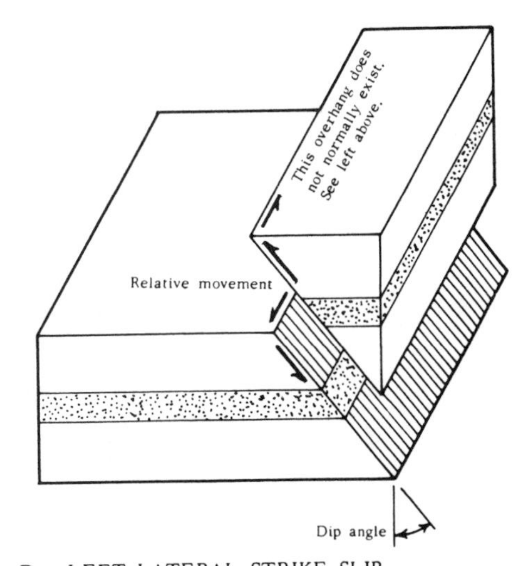

Figure 1-3 Types of earthquake faults. *Source:* K. V. Steinbrugge, private collection. Reprinted with permission.

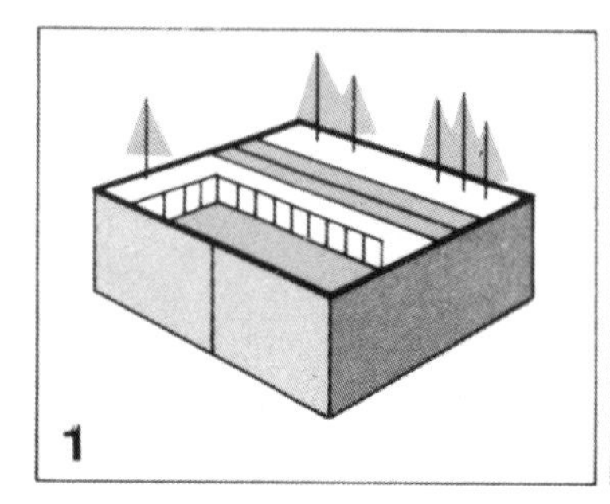

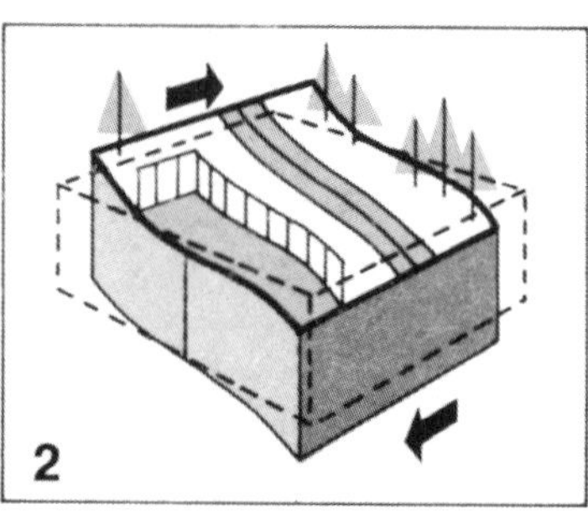

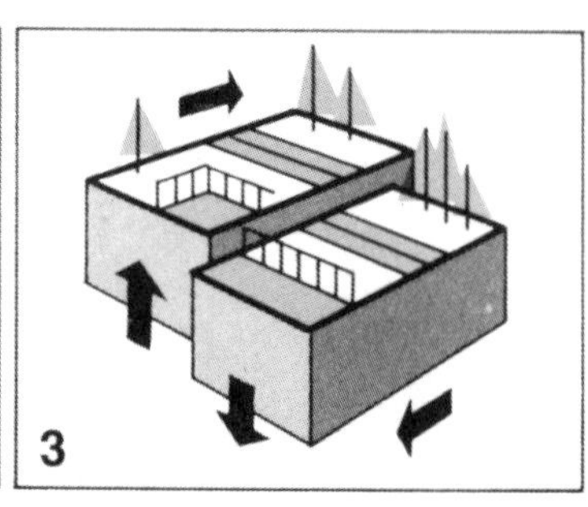

Figure 1-4 Diagram of movements associated with elastic rebound theory indicating preevent and postevent characteristics. *Source:* U.S. Geological Survey.

In general, Reid's elastic rebound theory has been accepted as an earthquake mechanism. It has been verified under several circumstances, and over the years has required only minor modification. (See Figure 1-4.)

Dr. Bruce Bolt, University of California at Berkeley, (Bolt, et al. 1977), indicates that

> the strain slowly accumulating in the crust builds a reservoir of elastic energy, just as, for example, a coiled spring, so that in some place, the focal point, within the strained zone, rupture suddenly commences, and spreads in all directions along the fault surface in a series of erratic movements due to the uneven strengths of rock along the tear. This uneven propagation of the dislocation leads to bursts of high-frequency waves which travel into the Earth to produce the seismic shaking that causes damage to buildings.

He concludes by indicating that "ground shaking away from the fault consists of all types of wave vibrations with different frequencies and amplitudes."

The Theory of Dilatancy

Another model used to explain the earthquake mechanism is the dilatancy model, which also relates to the fracture of crustal rocks. It becomes a compelling explanation, since it has been observed that when crustal rocks experience sustained stress, localized cracking may occur and the volume of rock swells or, technically stated, "dilates" in response. Where there is a presence of water as a potential lubricant, a mechanism for sudden rupture is provided within the crack configurations.

Ordinarily, it is apparent that the weight of the overlying rock (lithostatic pressure) already equals the strength of uncracked rocks at the crustal depth of about 5 kilometers with temperatures (around 500°C) and pressures appropriate for that depth. Under such conditions the rock would deform plastically, since the shearing forces required to bring about sudden brittle failure and slipping along a crack would never be realized if no other ingredient were introduced. However, the presence of water provides the mechanism for sudden rupture by reducing the effective frictional component provided by the characteristics of the crack boundaries and resulting dilatancy. Professor Bolt (1977) further indicates that "the full picture of the dilatancy theory of earthquake genesis is not yet clear, but the hypothesis is attractive in that it is consistent with precursory changes in ground levels, electrical conductivity, and other physical properties which have been noted in the past before earthquakes."

HOW EARTHQUAKES ARE MEASURED

Many attempts have been made to develop methods to measure the size and effects of earthquakes. Of all those developed, only two methods have been generally accepted and are still in use today: one used for measuring the size and pinpointing the location, or "magnitude" and "focus," of an earthquake; and another used for measuring the effect, or "intensity," of an earthquake. It is interesting to note that the two measure different characteristics of an earthquake, yet when used together, give a good picture of where the seismic event took place, how large it was, and what its impacts were on the built environment. In order to give a general idea of the size and physical effects of an earthquake, it is necessary to use both measures.

Magnitude Scale

First used in 1935, the Richter magnitude scale was named after its inventor, Professor Charles Richter of the California Institute of Technology in Pasadena. Still extensively used today to assign a sense of the size of an earthquake, it can also indicate the amount, or "magnitude," of energy released at its source, or "focus." A correlation between the Richter scale and the amount of total energy released has been derived, with a one-unit increase in magnitude approximating a 30-fold increase in energy. Thus, a three-unit increase in magnitude, from 5 to 8 for example, renders approximately a 30 × 30 × 30, or a 27,000-fold, increase in energy released. (See Figure 1-5.)

Recently, attempts have been made to supplement the Richter magnitude scale with a composite magnitude scale M_s that takes into account the characteristics of all seismic waves. The Richter scale is based on seismographic measurement of an earthquake's largest seismic wave. However, since earthquakes release many kinds of seismic waves, it is the intent of the recently developed

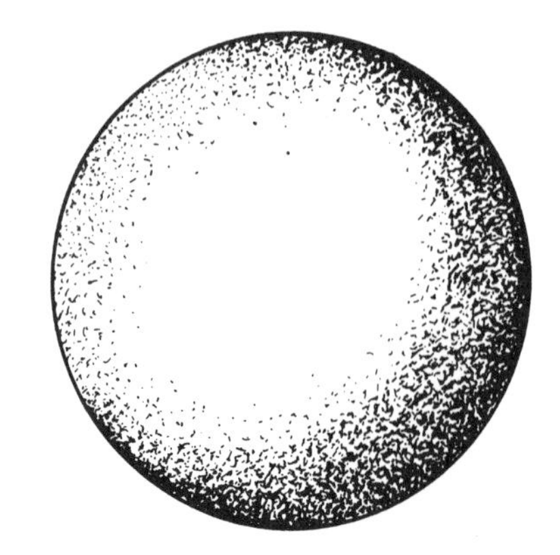

Figure 1-5 Comparison between earthquake magnitude and its energy release. The volume of the small 0.25-in. sphere is assumed to be the equivalent of Richter magnitude 1. A magnitude 2 earthquake would be a sphere 1.5 in. in diameter; a magnitude 3 about 5 in. in diameter; the 6.4 magnitude 1971 San Fernando earthquake would be a sphere 10 ft in diameter; and the 8.3 magnitude 1906 San Francisco earthquake a sphere of 93 ft in diameter. *Source:* K. V. Steinbrugge, private collection. Reprinted with permission.

comprehensive scale to measure the amplitudes of more of them. This explains why different magnitudes are often reported for the same event. For example, the October 1989 Loma Prieta earthquake was assigned a 7.0 magnitude on the Richter scale and a 7.1 magnitude M_s on the surface-wave characteristics.

The Richter scale reading is said to be an abstract number because it has no direct physical meaning, but rather "is intended to be a rating of a given earthquake independent of its place of observation." Because the scale was originally defined as the logarithm of the maximum amplitude recorded on a seismograph, every upward step of one magnitude unit represents the multiplication of the recorded amplitude by a factor of 10. Consequently, a Richter magnitude of 5 records 10 times the amplitude of a "Richter 4," and a "Richter 6" 100 times that of the "Richter 4."

On paper the scale is properly expressed in ordinary Arabic numerals and decimals. Although it is an open-ended scale with no upper limit, the largest known earthquakes have been those approaching a Richter 9.0. It is important to (1) remember when using the Richter scale that it is simply intended to give a numerical rating of a specific earthquake event and location in comparison to another, and (2) understand which magnitude scale is being used as a measure of the event.

Because of the abstract nature of the Richter scale, earthquakes of similar Richter magnitudes may differ greatly from each other in the physical effects produced on the built environment because of the immense variety of local geological conditions that affect the waves traveling through the earth and the fact that the Richter scale does not differentiate between deep-focused and shallow-focused earthquakes. Earthquakes generate two fundamental types of seismic waves: those that come from within the earth's interior (body waves) and those that travel along the the earth's surface (surface waves). Damages from all earthquakes result from (1) the depths at which the strongest waves initially fan out in all directions, and (2) the geological characteristics through which they must travel. Accordingly, the new composite magnitude scale M_s proposes to take into account the characteristics of all waves, whether of deep or shallow focus, and provide a more complete understanding of wave amplitude and behavior.

Intensity Scale

While the Richter magnitude scale indicates the size of an earthquake, it does not given an idea of the physical effects of an earthquake on buildings. We must therefore turn to a second scale to measure the intensity of local damage to structures and facilities. The important thing to remember is that we need both scales, one to determine an event's magnitude and epicenter and another to indicate its intensity. In order to obtain a more complete picture of the earthquake's size, location, and physical impact on buildings and facilities in the stricken area, both scales must be used.

Although a number of scales have been developed to describe the effects of ground shaking on the performance of buildings at a given location, the modified Mercalli intensity scale (MMI) is the most widely accepted in the United States. In contrast to the Richter magnitude scale, which uses Arabic numerals, the MMI utilizes Roman numerals, ranging from MMI-I to MMI-XII. This also avoids confusion in distinguishing between the two scales: an Arabic numeral means we are dealing with the Richter magnitude scale, and a Roman numeral indicates the use of the modified Mercalli scale as a measure of the relative amount of damage incurred.

Another difference between the two scales is that while the Richter scale is open-ended with no theoretical upper limit, the modified Mercalli scale is a closed-ended measure, with the maximum intensity of XII used to indicate "damage nearly total, the ultimate catastrophe." At the other end of the scale, an area of damage that has been assigned an intensity of MMI-I is described as: "Earthquake shaking not felt. But people may observe marginal effects of large distance earthquakes without identifying these effects as earthquake caused. Among them: trees, structures, liquids, bodies of water sway slowly, or doors swing slowly." Table 1-1 gives complete descriptions of each of the intensities, from I to XII, used in this scale.

When a damaging earthquake occurs, trained observers are sent into the field to assess overall destruction in the entire region. They assign intensity levels according to their evaluation of the damage

TABLE 1-1 Modified Mercalli Intensity Scale of 1931 (Abridged)[a]

I. Not felt except by a very few under especially favorable circumstances (RF I)

II. Felt only by a few persons at rest, especially on upper floors of buildings. Delicately suspended objects may swing (RF I–II)

III. Felt noticeably indoors, especially on upper floors of buildings, but many people do not recognize it as an earthquake. Standing motor cars may rock slightly. Vibration like passing of truck. Duration estimated (RF III)

IV. During the day felt by many, felt outdoors by few. At night some awakened. Dishes, windows, doors disturbed; walls make creaking sound. Sensation like heavy truck striking building. Standing motor cars rocked noticeably (RF IV–V)

V. Felt by nearly everyone; many awakened. Some dishes, windows, etc. broken; a few instances of cracked plaster; unstable objects overturned. Disturbance of trees, poles, and other tall objects sometimes noticed. Pendulum clocks may stop (RF V–VI)

VI Felt by all; many frightened and run outdoors. Some heavy furniture moved; a few instances of fallen plaster or damaged chimneys. Damage slight (RF VI–VII)

VII Everybody runs outdoors. Damage negligible in buildings of good design and construction; slight to moderate in well-built ordinary structures; considerable in poorly built or badly designed structures; some chimneys broken. Noticed by persons driving motor cars (RF VIII minus)

VIII. Damage slight in specially designed structures; considerable in ordinary substantial buildings with partial collapse; great in poorly built structures. Panel walls thrown out of frame structures. Fall of chimney, factory stacks, columns, monuments, walls. Heavy furniture overturned. Sand and mud ejected in small amounts. Changes in well water. Disturbs persons driving motor cars (RF VIII plus to IX minus)

IX Damage considerable in specially designed structures; well-designed frame structures thrown out of plumb; damage great in substantial buildings, with partial collapse. Buildings shifted off foundations. Ground cracked conspicuously. Underground pipes broken (RF IX plus)

X. Some well-built wooden structures destroyed; most masonry and frame structures destroyed with foundations; ground badly cracked. Rails bent. Landslides considerable from river banks and steep slopes. Shifted sand and mud. Water splashed (slopped) over banks (RF X)

XI. Few, if any, (masonry) structures remain standing. Bridges destroyed. Broad fissures in ground. Underground pipe lines completely out of service. Earth slumps and land slips in soft ground. Rails bent greatly (No RF)

XII. Damage total. Waves seen on ground surfaces. Lines of sight and level distorted. Objects thrown upward into the air (No RF)

[a] The parenthetical RF listings indicate equivalent intensities on the Rossi-Forel Scale. This scale was in use at the time of the 1906 San Francisco shock and is found on the current isoseismal maps.

Source: Steinburgge (1982). Reprinted with permission.

that has taken place in accordance with the descriptions of damage states listed in the Modified Mercalli Scale. Such assessments by trained observers, who are experienced design professionals or building department officials, may be influenced by their own subjective reactions and the personal accounts of residents in the area at the time of the earthquake.

Based on these field observations of destruction after an earthquake, an isoseismal map is prepared to indicate intensity levels of damage assigned to areas around the earthquake's impacted region. The isoseismal map is produced by drawing a line that connects points of equal intensity of damage states of buildings and facilities located in the effected area. Except for localized pockets of intense destruction in specific areas as a result of geologic characteristics in the region, intensity maps typically show that levels of damage decrease with increasing distance from the epicentral region, due to the attenuation of earthquake energy with distance. Figures 1-6 through 1-9 indicate examples of isoseismal maps drawn after the following four U.S. earthquakes:

1. 1933 Long Beach, California,
2. 1965 Puget Sound, Washington,
3. 1980 Kentucky, and
4. 1987 Whittier-Narrows, California.

SEISMIC RISK MAPS

All 50 states in the United States are subject to seismic events to some degree or another, with some more prone to severe earthquakes, as indicated by Figure 1. In order to give design professionals and emergency response planners a perspective of the relative seismicity of respective regions in the United States, national seismic risk maps have been developed over the years. Generally speaking, these maps are drawn on the basis of the location of historic earthquakes, their magnitude and intensity, their probability of recurrence, and the frequency of events in the same region.

It is increasingly important for the architect to become thoroughly familiar with seismic risk maps as a document that indicates levels of relative earthquake hazards in the country. Seismic risk maps have been adopted by building code officials in many areas and have become official documents in respective building codes such as the Uniform Building Code (UBC) and others. Accordingly, architects will automatically find themselves dealing with seismic risk maps in their normal practice. Figure 1-10 shows the current seismic risk map including in the 1988 edition of the UBC. Figure 1-11 indicates one of the first maps, derived by the Applied Technology Council (ATC) in 1978, for use in establishing tentative pro-

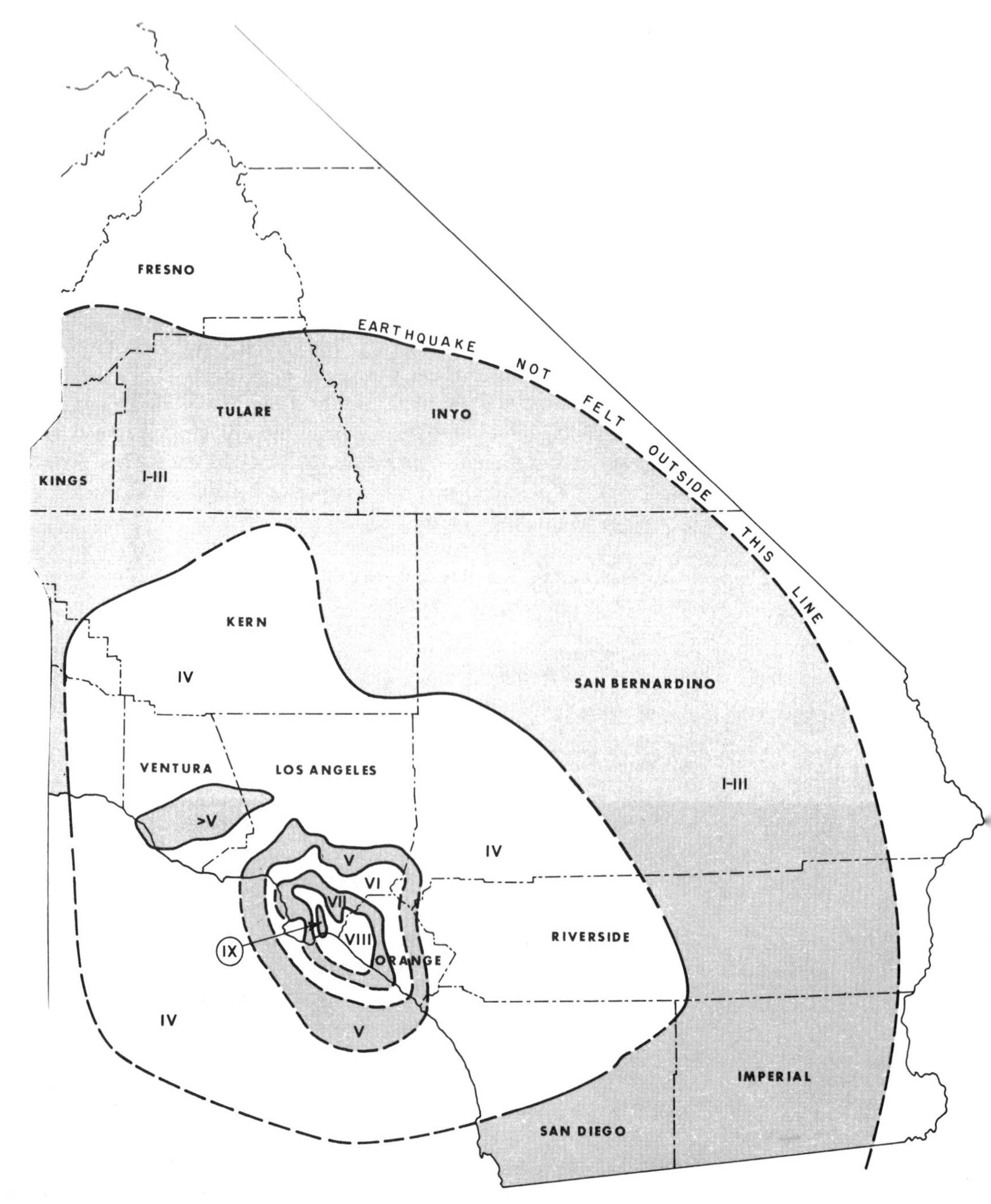

Figure 1-6 Isoseismal map of 1933 Long Beach, California, earthquake. *Source:* California Division of Mines and Geology.

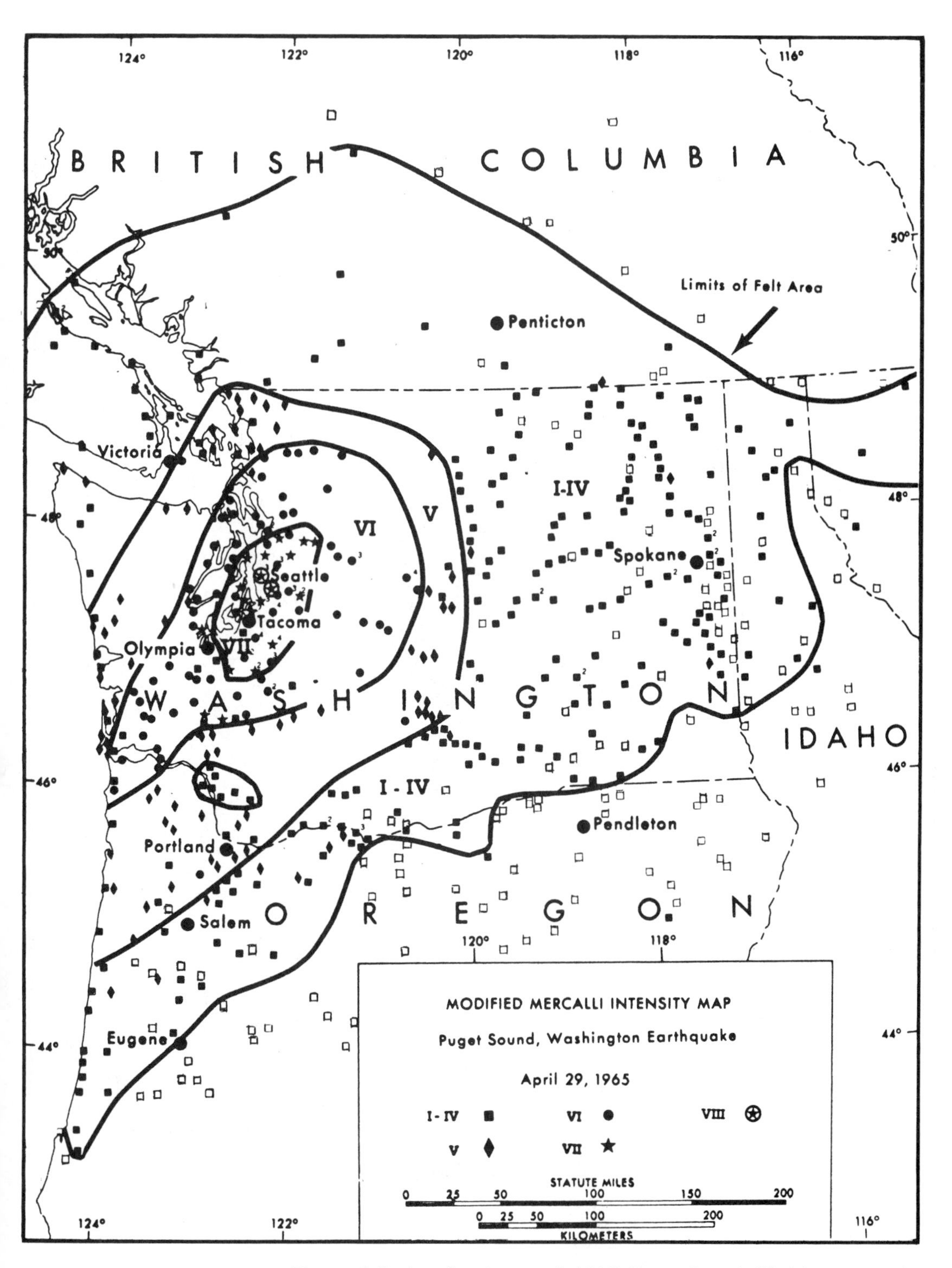

Figure 1-7 Isoseismal map of 1965, Puget Sound, Washington, earthquake. *Source:* K. V. Steinbrugge, private collection. Reprinted with permission.

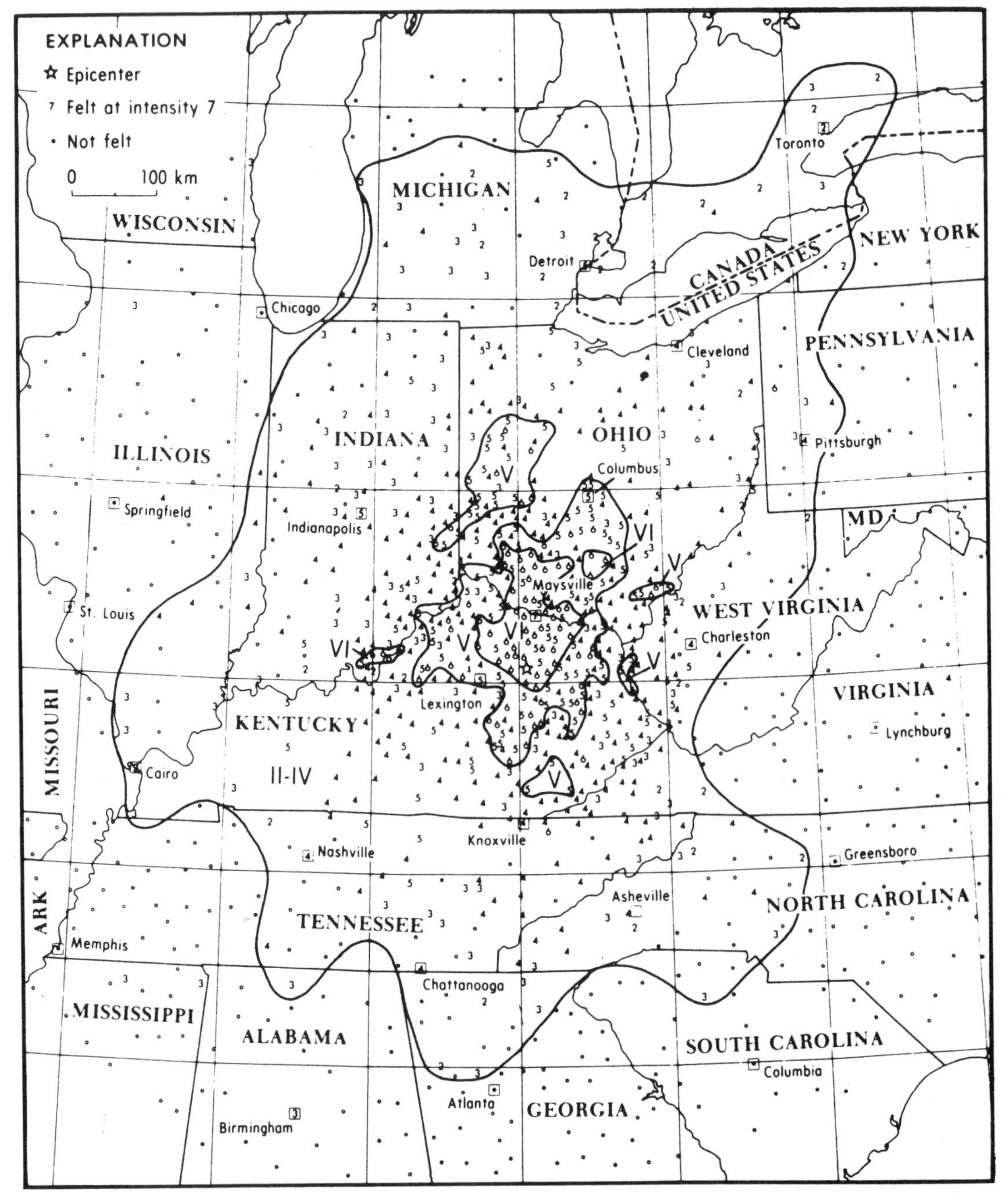

Figure 1-8 Isoseismal map of 1980 Kentucky earthquake. *Source:* K. V. Steinbrugge, private collection. Reprinted with permission.

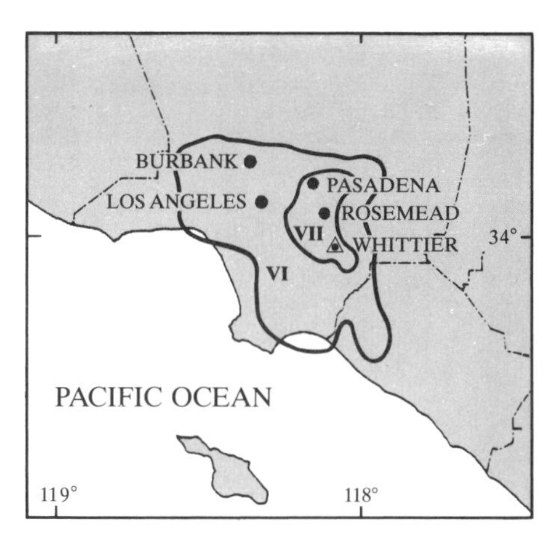

Figure 1-9 Isoseismal map of 1987 Whittier-Narrows, California, earthquake. *Source:* Eisner (1989a). Reprinted with permission.

visions for the development of seismic regulations for buildings on a national basis, which eventually led to the National Earthquake Hazards Reduction Program (NEHRP) provisions (see Chapter 10).

As can be seen from these two maps, the more seismically active areas in the United States are located in Washington, Utah, California, Nevada, Missouri, Alaska, Hawaii, and South Carolina. States with lesser incidence of earthquake activity are Texas and Florida. However, for the architect, it is clear that earthquakes are a national problem, and that the seismic performance of buildings designed according to current codes is becoming a very visible item on the architect's agenda. That seismic issues and concerns are now a national problem can be seen by the fact that the recently funded National Center for Earthquake Engineering Research (NCEER) is located in Buffalo, New York, and not California.

In current practice, it is rather common for larger architectural offices, as they expand into national and international markets, to become involved with design projects dealing with seismic forces. For this reason, whether or not the architect's office is located in a low seismic risk area like Texas or Florida is immaterial. What is important is that the architect be familiar with seismic risk maps and earthquake-resistant design performance standards in order to be competitive in the marketplace, whether in the United States or abroad.

LIFE SAFETY ISSUES

As mentioned in the introduction to this book, it is often said that "earthquakes do not kill people, buildings do." Historically, this aphorism has proven to be quite true, for it is clear that the poor

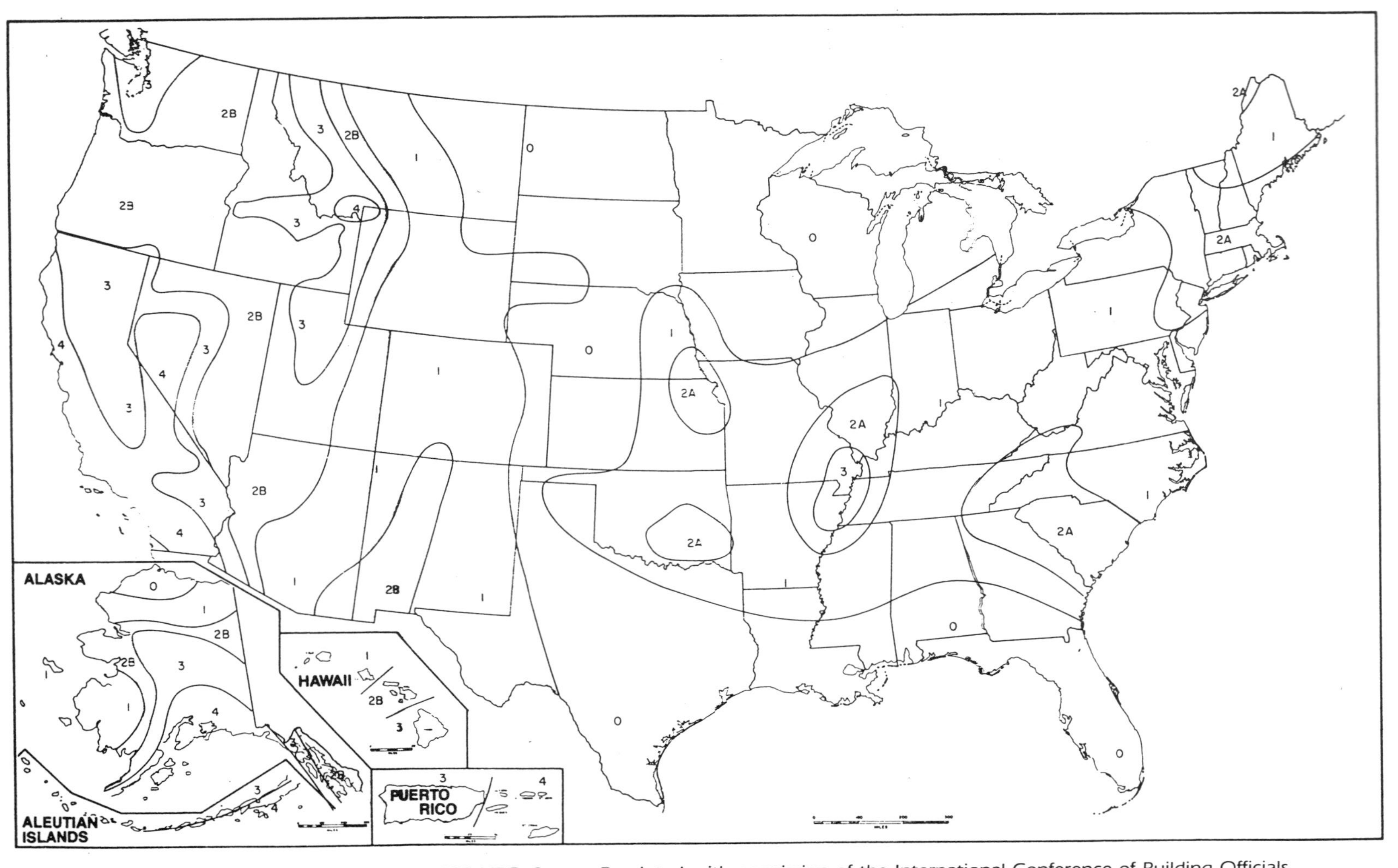

Figure 1-10 U.S. seismic risk map, 1988 UBC. *Source:* Reprinted with permission of the International Conference of Building Officials.

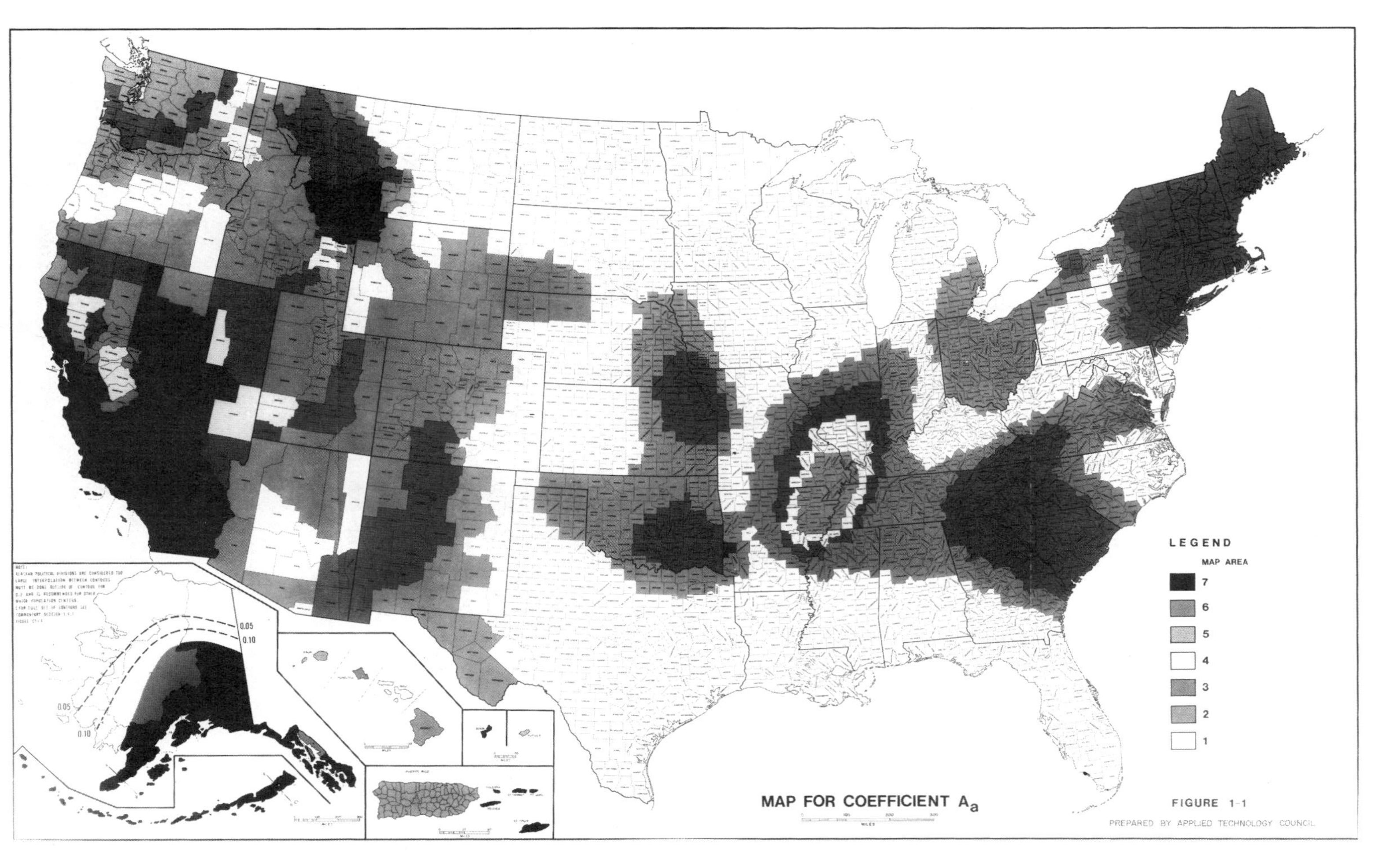

Figure 1-11 U.S. seismic risk map based on effective peak accelerations, 1978 ATC. *Source:* Applied Technology Council, Reprinted with permission.

seismic performance of buildings, leading to building collapse or severe damage, has been the principal cause of loss of life. The safest place to be during an earthquake is in an open field away from trees, telephone poles, electric transmission lines, man-made structures or facilities, or anything else of substantial weight or with a sharp edge that can fall on a potential victim. Even nonstructural building elements, suspended light fixtures, or precast exterior cladding components, for example, have been known to be the cause of life loss or injury during an earthquake. (For a more detailed presentation on the importance of nonstructural elements to seismic safety, refer to Chapter 6.)

In today's world of extreme exposure to professional liability, architects must be aware of their responsibilities in seismic safety design. Realistically, it will no longer suffice to stand back and ignore the problem by professing ignorance, or duck the issue by stating that it is an engineering problem after having passed the responsibility on to the consulting structural engineer. As the leader of the design team, the architect will be held accountable whether or not there has been direct involvement in the seismic design provisions of the building. The architect must act prudently by being sufficiently knowledgeable of seismic safety principles and objectives in order to justify preliminary decisions made on earthquake-resistant design concepts derived by the entire team.

In the seismic design of buildings, the architect must realize that life safety and damage control are the shared responsibilities of all design professionals. Owing to the growing complexity of our contemporary buildings and their dynamic, nonlinear performance when responding to major seismic events, these responsibilities must be shared with all parties involved in the planning, design, and construction of earthquake-resistant building systems. The seismologist, geophysicist, consulting engineers, architect, planner, building official, contractor/builder, and the owner/user must join together as a working team and not be concerned about "protecting turf." When dealing with the seismic design of buildings and facilities, things are just too complex to be handled unilaterally by just one individual.

VULNERABILITY AND RISK ANALYSIS

Generally, when determining the seismic risk involved for a given building, four factors must be considered: (1) hazard, (2) exposure, (3) vulnerability, and (4) location. Hazard encompasses all possible geologic hazards, such as ground-shaking potential, fault rupture, liquefaction, landslides, and tsunami, among others. Exposure refers to public health and safety in the face of the hazard. It includes the occupancy type, use, and function of a building. Vulnerability is associated with the potential survival capability or expected per-

formance level of the building system. Location deals with the proximity of the building to a potential earthquake source. To identify structures in the high-risk range, all four factors must be considered and assessed as having a positive or negative effect on the potential performance of a total building system.

A building defined as having an extreme vulnerability to earthquake hazards, owing to its construction type, for example, may not be placed in the high-risk category if it is located in an area not even remotely exposed to an earthquake source. In addition, it should be realized that even highly vulnerable structures do not automatically produce a high-risk/life-loss situation, since life loss is also associated with the type of occupancy and use of the building. A classic example used to illustrate this point is the case of a building used as a warehouse. If such a building, which is empty at night and has low occupancy levels even during working hours, is vulnerable to severe damage in an earthquake, it is still potentially less of a life safety risk to its occupants than a different building type less vulnerable to severe damage, such as a well-designed hospital with high occupancy levels of patients on life-support systems for 24 hours a day. Obviously, total risk in terms of potential life loss and injury increases with the number of occupants in a building. Theaters, movie houses, and other enclosed places of concentrated public assembly are therefore categorized at a higher level of risk in contrast to building types with less occupants.

Relative Earthquake Safety of Building Construction Types

Current seismic provisions in building codes are intended to protect life and reduce, not eliminate, the potential for minor property damage as long as it is not life-threatening. Even though we cannot yet precisely predict earthquakes with respect to specific time, location, and magnitude, it is apparent from past experience that the existing building stock in metropolitan centers located in areas of high seismic risk will be subjected to a major earthquake at one time or another. The potential of death or injury to people living or working in potentially hazardous building types is a major concern. By establishing the year 1933 as an applicable base in the development of earthquake-resistant design, it is possible to derive a vulnerability scale of building construction types based on historic experiences without consideration of external geological effects such as landslides, liquefaction, subsidence, flooding, or tsunami. Focusing on building records of past seismic performances alone, without the consequential effects of those external geological impacts, can provide the basis for an abstract technical analysis of the maximum probable deaths per building construction type per 10,000 occupants due to earthquakes.

Experience provides us with a relationship between building

construction type and life loss in that damage assessments are directly related to a building's anticipated level of performance. For example, worldwide records kept over decades of damaging seismic events demonstrate that 80 percent of the deaths related to earthquake hazards are attributable to unreinforced adobe and other low-strength masonry building construction types. In contrast, historic records also reveal that other construction types, such as one-story lightweight wood-frame buildings (without heavy roof loads), will survive earthquakes quite well, as they rarely experience total "pancake" collapse. Table 7-1 presents the results of an analysis made on the relative earthquake safety of representative building construction types based on their past performance levels.

It is necessary to realize that the figures listed in Table 7-1 indicate general projections of potential life-loss data based on a simplified classification of typical construction systems generally found in the existing building stock of our urban areas and are divorced from the geological effects mentioned earlier. In this regard, the figures shown are only useful as a general measure in assessing the relative safety of representative construction types in the form of a life safety ratio based on units of an abstract 10,000 occupancy rating.

In summary, therefore, when analyzing the relative safety of a building's design, it is essential to recognize four fundamental considerations: (1) current practice in seismic provisions of a typical building code does not anticipate producing "earthquake-proof" buildings with cost-effective construction techniques; (2) projected lower bound limits for life loss are estimated at the performance level of small, one-story lightweight, detached wood-frame buildings; (3) projected upper bound limits for life loss are estimated at the performance levels of unreinforced adobe and other low-strength masonry structures; and (4) a building's age, size, type of construction, and quality of maintenance over the years have a direct bearing on its expected seismic performance level and relative safety.

CHAPTER

2 GENERAL ASPECTS OF BUILDING PERFORMANCE

A second set of questions is often triggered when the architect starts to discusss the seismic performance of a new building or the remodeling of an existing building with the client:

"How safe will my building be when a major earthquake takes place in the immediate vicinity?"

"Will performance standards required by the building code guarantee the safety of occupants in my high-rise apartment complex located in a region of high seismic risk?"

"When an earthquake occurs, what will be the level of damage, if any, that should be anticipated in my old masonry building after we remodel it?"

These are good questions to be asked, but again, unfortunately, there aren't any simple answers because of the many variables involved. In earthquake-resistant design, the approach to be taken depends on many factors, each of which must be given careful consideration. Lamentably, there are no easy "cookbook" recipes available. These questions are also germane to discussions when a client, seeking to purchase an existing structure, asks the architect for a comprehensive seismic evaluation of the building and its site.

It is crucial that architects objectively discuss these questions with all clients even before starting the preliminary design phase of any new building so that a common understanding of what is expected is reached. In terms of the building code it is important that clients understand that codes specify *Minimum* standards to be followed, not optimum ones. With all the probabilistic uncertanties involved in seismic design, it is essential that the architect act prudently with the client by clearly defining the limits of earthquake-resistant design before any preliminary drawings are initiated.

The actual effects that earthquakes have on buildings depend on many uncertainties, including (1) the magnitude of the earthquake, (2) geological characteristics of the site, (3) location of the earthquake's focal point, (4) severity and duration of the ground shaking, (5) the fundamental period of the ground motion waves according to soil types, (6) code provisions in force at the time of the building's design, (7) the building's construction type, (8) the building's design and configuration, (9) quality of construction, (10) and proper building maintenance by the owner once construction is finished. One of the most important variables used to evaluate a building is its date of construction, since this determines the edition of the building code that governed design and construction procedures at the time. Since building code provisions are normally not retroactive, many buildings still in use today may have been designed according to older, outmoded standards, depending on their age. Thus, one may argue that given the age of a building, an experienced design professional would be able to forecast its expected performance during a major, damaging earthquake quite accurately.

A good example of this is found in the collapse of the Highway I-880 Cypress Street double-decker overpass during the October 17, 1989, Loma Prieta earthquake in the Santa Cruz Mountains of California. (See Figures 12-5, 12-6, and 12-7.) The ill-fated overpass was built in 1958 under design standards then in effect but now considered outmoded. Upon evaluation of the pancake failure of its upper deck, which led to the deaths of 41 persons, it was clear that design standards for reinforced concrete structures had changed drastically since 1958. Indications are clear that it would not have collapsed had the 1988–1989 standards now in effect been followed.

THE BUILDING CODE PROCESS FOR SEISMIC REQUIREMENTS

One of the first building codes to include seismic provisions was the edition of the Uniform Building Code (UBC) issued after the major, damaging 1933 Long Beach earthquake in which many unreinforced masonry buildings collapsed or suffered severe damage (see also the section on the California Field Act of 1933 in Chapter 10). Prior to that year, seismic (lateral load) code provisions were practically nonexistent. Consequently, 1933 becomes a key date in assessing the anticipated seismic performance of an existing building. Based on this date, a valid assumption could be made that most structures designed and built before 1934 in California had little or no earthquake-resistant design provisions. And, in general, such buildings would be expected to perform poorly if required to withstand major lateral forces, depending on the magnitude of the earthquake-induced forces, of course.

Because new lessons are learned about the seismic performance of buildings after each major earthquake, it is not uncommon for building codes to have their seismic provisions updated after each damaging event. As part of this process, damage patterns are observed and information is collected in the field by reconnaissance teams of five or more professionals immediately after a major earthquake. All professional disciplines are represented in such reconnaissance efforts. Over the years, these preliminary field studies have been called "earthquake reconnaissance reports." The information reported is then carefully studied and appropriately evaluated for further research. In the final step, the findings resulting from the detailed research studies that follow are then fed back into the process by which code provisions are modified and new standards promulgated.

Because of the time needed to confirm and verify findings, there is considerable time lag in this cycle of introducing new data into the latest editions of the code. Architects should be aware of this cycle and realize that some of the latest research findings have yet to be introduced into the building code. There is one exception to this time lag: the swift introduction of "emergency" seismic code provisions during the immediate response phase that follows a particularly damaging earthquake. In such cases, when previous standards, or force coefficients used for building design purposes, are found to be insufficient to resist the lateral forces generated by the earthquake that has just occurred, prompt action must be taken. Emergency code provisions, for example, were issued as soon as possible to increase the ductility coefficient for immediate use in the design of any new multistory buildings in Mexico City after the 1985 Mexico/Michoacán earthquake.

HOW SEISMIC-INDUCED LOADS ARE TRANSFERRED TO BUILDINGS

An important fundamental for architects to understand from the very beginning is how earthquake-induced forces are translated into the building. When any rupture occurs in an earthquake fault zone, it produces a multitude of vibrations or seismic waves that emanate in all directions. Although the focus of the earthquake is identified as the initial point of energy release and source of rupture, the actual surface faulting that follows may spread in a predominant direction for many miles from this central spot. In any event, the result is a random pattern of ground-shaking motions that is felt throughout the entire impacted area.

The effects of this ground shaking may be very severe on a building in terms of the motion's amplitude, velocity, acceleration, displacement, and duration, depending on the magnitude of the earthquake. Essentially, this ground shaking is caused by the seismic body and surface waves discussed earlier that are propagated in

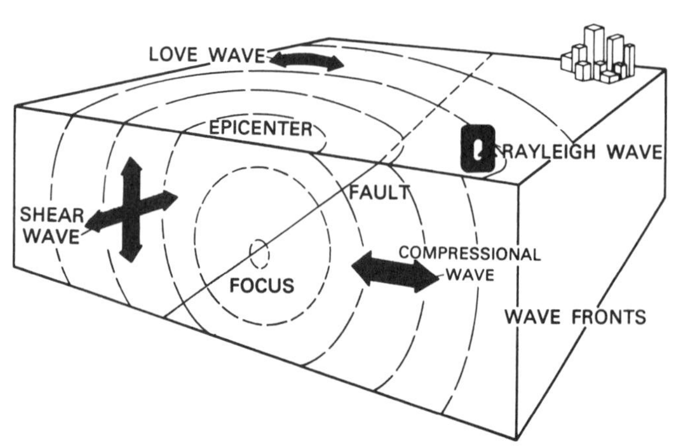

Figure 2-1 An earthquake produces a set of primary seismic waves that are propagated in all directions and produce a random pattern of ground-shaking motions. *Source:* U.S. Geological Survey.

all directions (see Figure 2-1). These motions are translated into dynamic loads that cause the ground, and consequently any buildings located on that shaking ground, to vibrate in a very complex manner. The result is that the structure is subjected to a combination of horizontal and vertical loads introduced through the foundation system fixed to the ground. Of the two loads involved, the horizontal one is the greater by far, although the vertical forces are currently under considerable study.

Another way of looking at the manner in which the seismic forces are translated to the structure is to realize that whereas the building foundation is fixed to the ground, thereby having a general tendency to be displaced along with the site, the upper portion of the building, or superstructure, is not; it is therefore free to move horizontally or oscillate at its own natural period. In tall buildings, depending on which mode of vibration the building enters, some floors may be moving in one direction while others may move in another at the upper stories. Interstory drift between floors may occur, leading to a foreshortened distortion of the building's stories, as illustrated in Figure 2-2.

How the Sizes of Earthquake Forces Are Derived for Building Design Purposes

As the base of the building responds to the ground motions produced by an earthquake, the bottom of the structure moves immediately, while the upper portion tends to lag behind due to its mass of inertia. The total lateral seismic force that results and that must be laterally resisted by the building system, by convention, is represented by the symbol V. The lateral force V, often referred to as the base shear, is equal to the mass M, or total weight of the building,

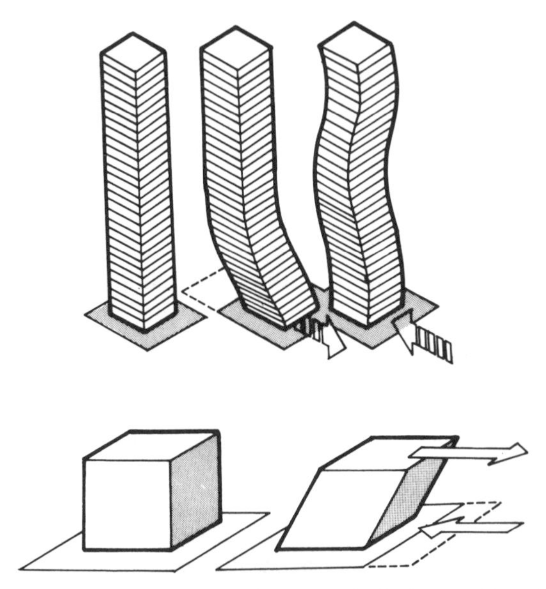

Figure 2-2 Drift diagrams indicating lateral displacement and resulting foreshortening of a building system. In relatively tall structures, some floors tend to move in one direction while floors above and below may move in opposite directions. *Source:* Botsai, et al. (1976). Reprinted with permission.

multiplied by the acceleration A produced by the relative earthquake motions. This equation may be written as V (total lateral force) = M (total building mass) × A (acceleration produced by ground motions), or

$$V = MA.$$

The equation displayed above will be recognized as a derivation of Sir Isaac Newton's work, which formulated and proved the law of gravity as a product of mass and velocity. For building code purposes, which make allowances for acceleration, damping effects, and other perceived structural response characteristics, the formula to determine the total lateral forces used for building design may be rewritten in abbreviated form as

$$V = CW,$$

wherein:

V = base shear (or total lateral load to be resisted).
C = seismic coefficient (written as a factor modifying acceleration and structural response characteristics).
W = total mass, or weight, of the building system.

Accordingly, it is quickly realized that the heavier the mass *M* (or *W*), as the total weight of the entire building, and/or the higher acceleration *A* (or *C* factor) of the earthquake motions, the greater the resulting force *V*, base shear, to be applied to the design of the building system.

Eventually, the resulting loads that act on buildings may be translated into shear, compression, tension, or torsional (rotational) forces. To avoid failure, the total building system must be designed with the possibility that all four forces will be present in one way or another, or in any combination, during an earthquake, depending on its magnitude. In approaching the seismic design of a structure, as an initial point of departure, it is always essential to examine the entire building system closely for potential exposure to these four forces or the effects of their combination.

HOW SEISMIC-INDUCED FORCES ARE RESISTED IN BUILDINGS

Ordinarily, if dynamic lateral forces, such as those caused by earthquake or wind, were not taken into consideration, the structural design of a building system would be a relatively simple process, since only vertical gravity forces need be taken into account. In such instances, all forces produced by dead loads and live loads are calculated in terms of gravity (1g) and applied vertically. As these loads are considered static, there is no need for structural components, technically speaking, to resist any dynamic lateral forces (see Figure 2-3*a*). However, in the case of earthquake loads acting on a building, it is a totally different situation (see Figure 2-3*b*).

Earthquake loads are dynamic and are applied laterally (horizontally) to a building system. When forces induced by an earthquake are spread throughout the buildings as horizontal forces, their accumulations are applied by convention to the structural system at each floor line and roof level (see Figure 2-3*c*). The rationale behind this convention is the assumption that it is through the floor and roof systems that horizontal forces are distributed throughout the rest of the building and eventually down through the vertical supports into the foundation.

Once the forces are applied to the floor and roof levels and distributed to other components of the basic structure, a vertical bracing system must then be designed to resist these forces and carry them throughout the rest of the building in a logical manner without breaking continuity in the integrity of the overall structural system. Basically then, we must design the vertical supporting system with the capacity to resist the lateral forces applied at the floor and roof levels throughout the building without excessive distortion and/or failure.

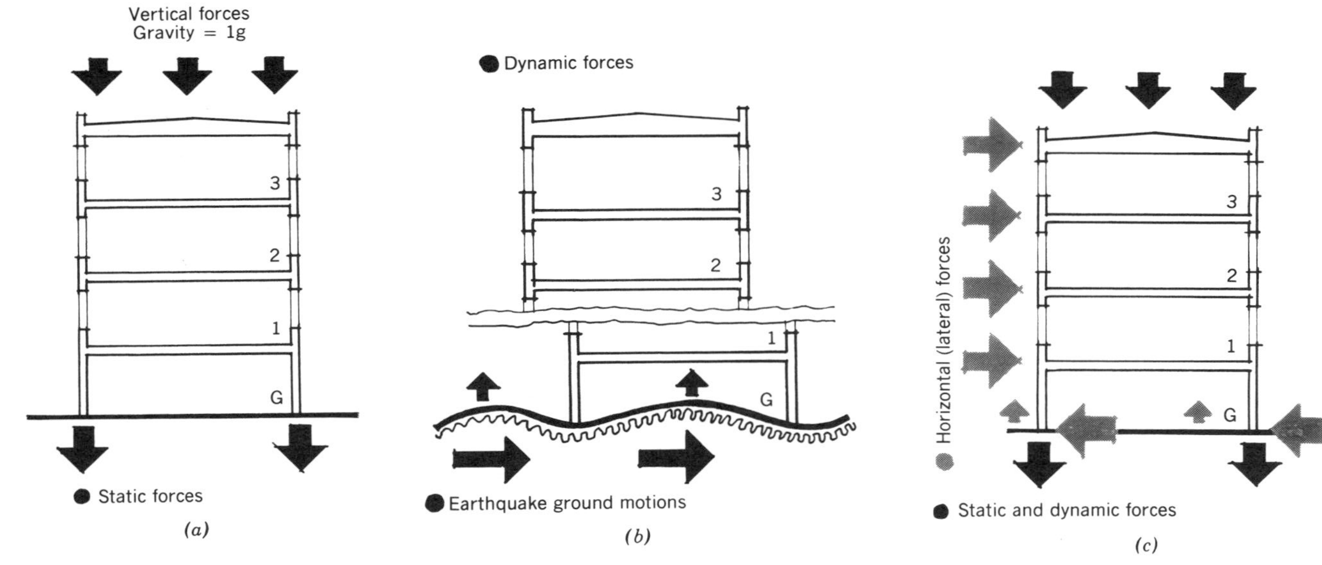

Figure 2-3 *(a)* Building system subjected to vertical gravity loads/static forces only; *(b)* building system subjected to earthquake-induced ground motions and dynamic lateral and vertical forces; *(c)* combination of gravity/static, dynamic lateral, and dynamic vertical loads applied to a building system.

An analogy to this is building a simple "house of cards." As we discovered as children, the house of cards was stable and remained intact as long as all the loads were predominantly vertical. However, we soon learned that if someone pushed horizontally against the house of cards with a lateral force larger than the vertical pull of gravity holding the cards together, the house collapsed, much to everyone's dismay. To take this analogy one step closer in terms of earthquake loads, we also discovered that if someone gave the table on which the house of cards was assembled a hard, sharp shove laterally, a similar collapse of the structure would take place. By visualizing the moving table as inducing motions into the house of cards, and noting its similarity to the dynamic ground shaking produced by an earthquake, we now have an excellent example of what type of action may occur to cause the collapse of a laterally unbraced structure impacted by a severe seismic event. In effect, we have produced a simple earthquake simulation model quite similar in basic principle to the engineering "shaking table" used for experimental research in the laboratory testing of structural building systems and components. The conclusion to be drawn from this "house of cards" analogy is that some type of lateral bracing system must be introduced into the basic structural system to avoid collapse or failure.

LATERAL BRACING SYSTEMS IN BUILDING DESIGN

Pin-connected square or rectangular shapes are not inherently stable in their ability to resist horizontal forces unless bracing elements are added into their basic configurations. On the other hand, pin-connected triangular shapes of any form or size are inherently stable and will resist any load from any lateral direction. The design problem, therefore, becomes one of adding triangular subelements into the square or rectangular system to make it stable against lateral forces or changing the pin-connected joint into another type. Because the majority of structures in the built environment are rectangular in nature, the addition of triangular subelements or fixed joints to the fundamental building system, rather than free-to-rotate pin-connected joints, has become the basic traditional approach to earthquake-resistant design.

After years of experience, three basic and cost-effective structural bracing systems have been developed for the earthquake safety of buildings:

1. Shear wall system (boxed walls),
2. Diagonal bracing system (braced frame), and
3. Moment-frame system (fixed joints).

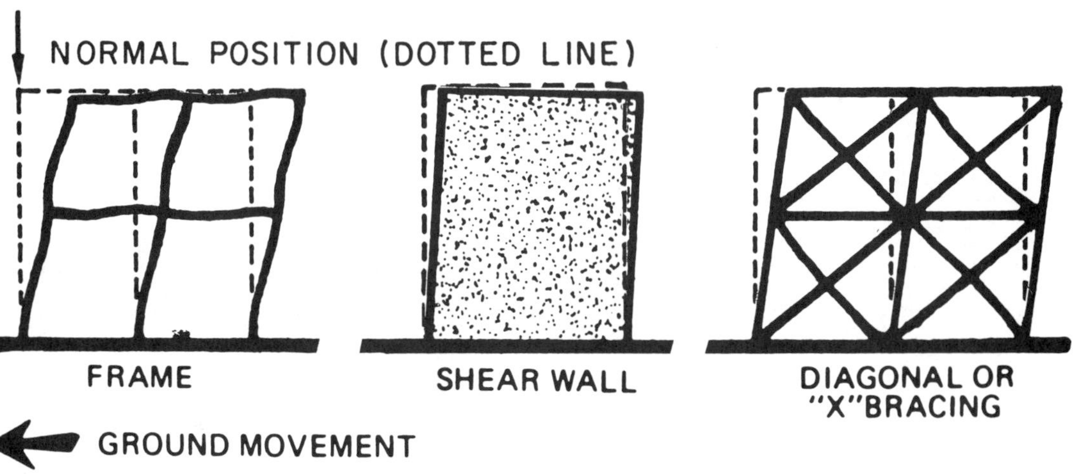

Figure 2-4 Three bracing systems most commonly used in the structural design of earthquake-resistant buildings.

Based on a traditional approach to seismic safety, these three systems are the most commonly used in the structural design of earthquake-resistant buildings today. Figure 2-4 illustrates their basic characteristics.

The relatively new base isolation approach, used to decouple the building system's superstructure from the ensuing ground motions and reduce the magnitude of earthquake forces induced through the foundations, will be presented in Chapter 5, "Building Design."

Building Classification and Construction Types

One of the simplest ways to identify and classify a building is to relate it to its basic structural and construction system and develop a description of its type. The classification of buildings is necessary for clear identification of damage patterns resulting from seismic forces and, consequently, to improve the performance of structures in future earthquakes. It is anticipated that all types of structures have different performance characteristics from one another depending on the construction materials and methods used to build them. Based on this performance, it is statistically possible to assess a building type's capacity to resist earthquake loads. From a building code point of view, this is important documentation to have available.

For the purposes of this book, a simplified classification system will be identified and used throughout the entire text. Although different from a more comprehensive system used by the insurance industry and another more complex one developed by the Applied

TABLE 2-1 Classification of Building Types

Class No.	Description
1A	Wood frame
1B	Heavy timber
2A	Small, light, all metal
2B	Large-scale, light, all metal
3A	Steel frame with steel decking floors
3B	Steel frame with reinforced concrete floors
4A	Reinforced concrete frame
4B	Precast reinforced concrete frame
4C	Tilt-up reinforced concrete system
5A	Reinforced masonry bearing wall
5B	Unreinforced masonry bearing wall

Technology Council (ATC), its simplicity and succinctness will be appreciated and more than meet the purposes of this publication. Table 2-1 lists the basic building types with which architects should be familiar.

As can be seen, the simplicity of the classification system presented in the table gives broad general categories of building types. It purposely does not go into extreme detail to describe subelements and subcategories of building classes, although such detail would be required by the insurance industry, which has other objectives tied to its classification system.

Building Damage Characteristics

The manner in which a structure absorbs, transfers, and counters the dynamic loads released by an earthquake determines whether it performs successfully (i.e., with little or no damage) or fails dramatically (i.e., with heavy damage). However, economic limitations have shown that it is absolute folly for an architect to promise a client an "earthquake-proof" building regardless of the earthquake's magnitude, as it has been seen that some damage, to one degree or another, will always occur to and/or in a building during a severe earthquake. Even a superficial plaster crack is classified as damage, as is a glass bottle of imported perfume that falls off a shelf and shatters on the floor.

The same problem has been faced in the area of mitigation of fire hazards in structures. The building industry has changed its philosophy by referring to "fire-retardant" materials or "fire-resistant" construction, and no longer advertises "fireproof" ratings as it once did many years ago. Accordingly, building performance standards clearly lean toward life safety over damage control by indicating that the code accepts the seismic design of buildings that "resist major earthquakes of the intensity or severity of the strongest experienced in California without collapse, but with some structural as well as nonstructural damage."

Exceptions to this philosophy do exist in the design of special service structures. For example, it is obvious for life safety reasons that nuclear power plants should be designed to much higher standards than other building types. Another example can be found in California, where, as mandated by the state in the 1972 Hospital Construction Act passed after the 1971 San Fernando earthquake, general acute hospitals are considered critical emergency service structures and are therefore designed to a higher standard to ensure that they remain operational after a major earthquake. For them to remain functional after a damaging seismic event, many of their nonstructural building elements are carefully braced and made earthquake resistant (refer to Chapter 10 for additional detail).

Building damage occurs when any component of a structure is loaded beyond its capacity to resist an applied force of any given magnitude. When any building element can deform and not compromise its ability to return to its original state without permanent deformation, the material has successfully maintained its integrity by staying within its elastic range. Any range of deformation that pushes a material beyond the elastic range into an inelastic range becomes very critical, because at that point the material will never be able to return to its original state. In such an instance, the component would be described as having failed, for once the elastic range is exceeded, permanent deformation or fracturing may also occur. This is particularly true for brittle materials, which under certain circumstances have a shorter elastic range and a tendency to fail abruptly, as compared to ductile components, which have the ability to remain more flexible through various cycles without total degradation while they continue to absorb energy. During the 1985 Mexico/Michoacán earthquake, many multistory buildings in Mexico City were subjected to as many as 40 complete cycles of extreme reversal due to the long duration of ground shaking, long period motions, and the effects of resonance. Hence, ductility and stiffness become vital characteristics for seismic safety, but they must be handled correctly and objectively.

Theoretically, therefore, to avoid failure an ideal building should be designed with unlimited capability in stiffness and/or with a capacity for boundless flexibility. However, most structures have a combination of some construction components that are brittle, or rigid, and others that are generally considered ductile, or flexible. Incorrect use of these components, whether they be structural or nonstructural, may be critical to the eventual seismic performance of a building system. The combination of stiff masonry infill walls, for example, when installed incorrectly between moment-resisting frame members, which are much more flexible, is often recognized as a source of typical damage patterns in buildings when the wall is not designed either as an effectively integral component of the frame or as one completely cut free from the frame's action.

When it comes to building damage, earthquakes have earned a growing reputation for their propensity to find the "weak link," no matter how small, in any complex system and exploiting it as a lead into progressive and total failure. As a result of this ability to search out and strike the weakest point of an assembly, it is often said that earthquakes "love" complexities. The more complex a building system, the more determined a major earthquake seems to be in quickly discovering and damaging the most vulnerable element. An important caveat in designing buildings to resist earthquake forces is therefore found in that well-quoted phrase, **"Keep it simple."**

BUILDING DAMAGE AND MAJOR EARTHQUAKES

Taking into account the potential magnitude of earthquakes and their inclination to seek out the weakest point in structures, the architect's preliminary planning and design decisions often have critical implications for damage patterns and life safety in building systems. On the one hand, with improved structural design methods and building standards, total building collapses are becoming less frequent; on the other hand, because of these improvements, nonstructural components have become increasingly exposed to damage. Although research study on the subject is still in progress, in calculating total dollar loss due to earthquake-induced damage, little differentiation is made between structural and nonstructural

TABLE 2-2 Damage Losses in Selected U.S. Earthquakes

Year	Location	Damage Loss (in Millions)
1906	San Francisco, California	$524.0
1918	Puerto Rico	$4.0
1933	Long Beach, California	$40.0
1935	Helena, Montana	$4.0
1946	Hawaii	$25.0
1949	Puget Sound, Washington	$25.0
1952	Tehachapi Mountains California	$50.0
1959	Hegben Lake, Montana	$11.0
1960	Hilo, Hawaii	$25.5
1964	Prince William Sound, Alaska	$500.0
1965	Puget Sound, Washington	$12.5
1971	San Fernando, California	$553.0
1987	Whittier-Narrows, California	$350.0
1989	Loma Prieta, California	$8,000.0

losses. Table 2-2 indicates examples of damage losses incurred in selected U.S. earthquakes after the turn of the century.

Damage Related to Ground Motion and Ground Failure

As indicated earlier, the physical results generated by a major earthquake manifest themselves in two ways: ground motion and ground rupture, both of which may be responsible for serious damage. Ground motions produced by an earthquake are the principal cause of building damage because they spread out in all directions to cover a large area, as shown by the isoseismal maps in Figures 1-6 through 1-9. Comparing the two hazards, ground rupture is not as extensive and therefore is less of a overall threat, although ground rupture has also led to severe building damage in specific situations.

Ground Failure

Ground failure during an earthquake may manifest itself in several ways, each of which may be a direct threat to buildings located in the immediate area. Principal examples of ground failures include surface rupture, landslides, liquefaction, and subsidence. A given earthquake may or may not produce ground rupture along a fault zone. If an actual rupture of the surface does occur, it may extend over several miles, as in the 1906 San Francisco earthquake, or it may be very limited. (See Figure 2-5.) Ground displacement along the fault can be horizontal, vertical, or both. The actual displacement may be measured in inches or in several feet, as the case may be. At times it has resulted in a surface break along a sharp, well-defined line, but usually tends to be distributed along a wider fracture zone.

Because there is a site-specific relationship between building location and damage in this case, the impact of ground failure on damage patterns as a distinct subject will be dealt with in detail in Chapters 3 and 4, "Site Investigation" and "Site Planning."

GROUND MOTION

By far, the greatest number of structures located in a metropolitan center will be impacted by earthquake hazardous ground motions. Examples of ground motions experienced during an earthquake are illustrated in Figure 2-6. When severe ground vibrations occur during major earthquakes, literally everything placed on the ground vibrates in response. Accordingly, it definitely becomes the principal consideration in the design of seismic-resistant buildings. If the ground vibrations and displacements are severe enough, they may

Figure 2-5 Accumulation of horizontal strike–slip (lateral) offsets over many years along San Andreas Fault Zone, California. *Source:* California Division of Mines and Geology (CDMG).

destroy any building not designed and constructed to earthquake-resistant standards.

An interesting aspect of the 1985 Michoacán earthquake is that major damage and life loss occurred in Mexico City, even though the city was located 200 miles from the epicenter. Approximately 5700 buildings located in the central downtown area of Mexico City were severely damaged or destroyed. This image of destruction tends to be counter to the public's normal interpretation of seismic events, wherein it is falsely expected that maximum damage always occurs in the epicentral region. This is one of the lessons to be understood by architects: Depending on the set of circumstances involved, distant earthquakes, as well as those nearby, may and do cause major damage in metropolitan centers whose location is not even remotely near the epicenter.

Long-Period/Short-Period Motions

It is implied that total earthquake energy dissipates with distance from the fault, but this is only generally true in specific cases and

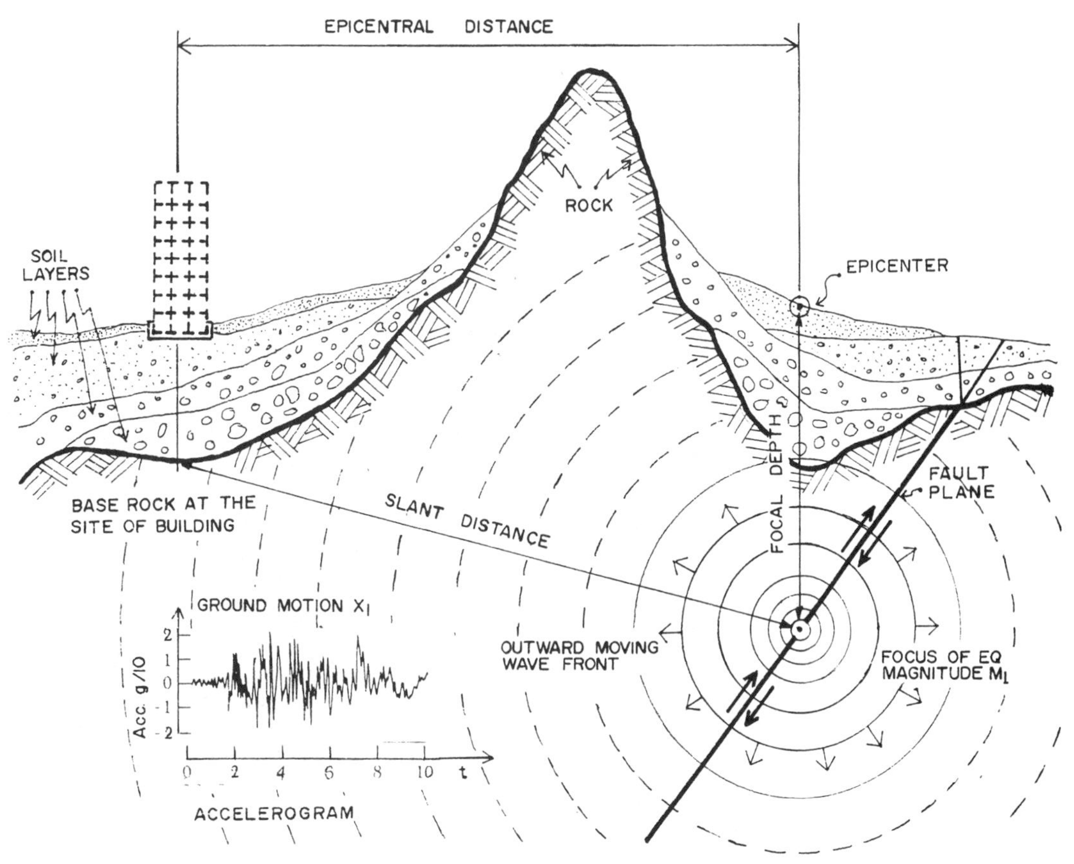

Figure 2-6 Focus of an earthquake and earthquake-induced ground motions. *Source:* Guevara (1989). Reprinted with permission.

may prove to be deceptive when not properly understood. It would be misleading to conclude that this automatically leads to less seismic risk to life and property located far from the fault zone.

Most damaging earthquakes throughout the world are associated with a relatively shallow depth of focus of less than 20 miles deep. The energy released by a shallow-focused earthquake may be spent over a small area, relatively speaking, and short-period ground motions do tend to die out more rapidly. In comparison, energy released from a deep-seated earthquake will travel greater distances and be felt over a larger region as a longer-period undulating motion. Generally, long-period motions are defined as those with cycles of 0.5 to 1 second or longer time periods.

Long-period motions tend to coincide with the longer natural periods of vibrations of deep unconsolidated soil pockets and multistory buildings over eight stories in height. When long-period motions coincide with the natural period of vibration of tall structures, resonance may occur with each successive cycle, causing larger and larger displacements at the upper stories of the building,

Figure 2-7 Illustrations of building damage after 1985 Mexico City earthquake.

depending on the duration of the earthquake motions. Low-rise buildings have short periods of vibrations and therefore do not tend to go into resonance with long-period waves. As an example of this manifestation in tall structures, the epicenter of the 1964 Alaska earthquake caused significant damage to multistory buildings in Anchorage, 75 miles away, whereas low, rigid structures in the same areas were not exposed to comparable damage.

In 1985 in Mexico, the recurrence of an earthquake similar to the 1957 Mexico earthquake, which had an epicenter located along

Figure 2-7 (Continued)

the coast 170 to 200 miles from Mexico City, produced long-period motions in resonance with the natural period of certain multistory buildings, causing many structures to collapse in the downtown area of Mexico City. The city's lakebed ground motions at the time were estimated to have a period of about 1.5 to 2 sec, which matched the natural period of buildings 9 to 14 stories in height. The results were disastrous for buildings of that size located on sites within the lakebed soils area. See Figure 2-7 for examples of damaged buildings in Mexico City. In both the 1957 and 1985 earthquakes, objects did not even fall from the shelves in low-rise structures located in Mexico City (see Figure 2-8).

The 1987 Whittier-Narrows earthquake in southern California had the opposite effect on buildings in the area: objects and merchandise in low-rise buildings were thrown from the shelves and many low, rigid structures were heavily damaged. (See Figure 2-9 for an example of a typical building damage pattern in downtown Whittier.) Its epicenter was only about four miles from the Whittier downtown area, which was mostly comprised of low, one- to two-story buildings typical of many cities in the Los Angeles basin.

Figure 2-8 Undisturbed objects on shelves in the Museum of Anthropology, Mexico City, 1985 earthquake.

ARCHITECTURAL AND PLANNING FACTORS INFLUENCING BUILDING PERFORMANCE

Building Period

In the preliminary design phase, it is very important that the architect understands the basic principle of potential resonance in a building system. It is also critical to know that a tall building system may have more than one natural period, or mode, of vibration de-

Figure 2-9 Unreinforced masonry building damage after 1987 earthquake in Whittier-Narrows, California.

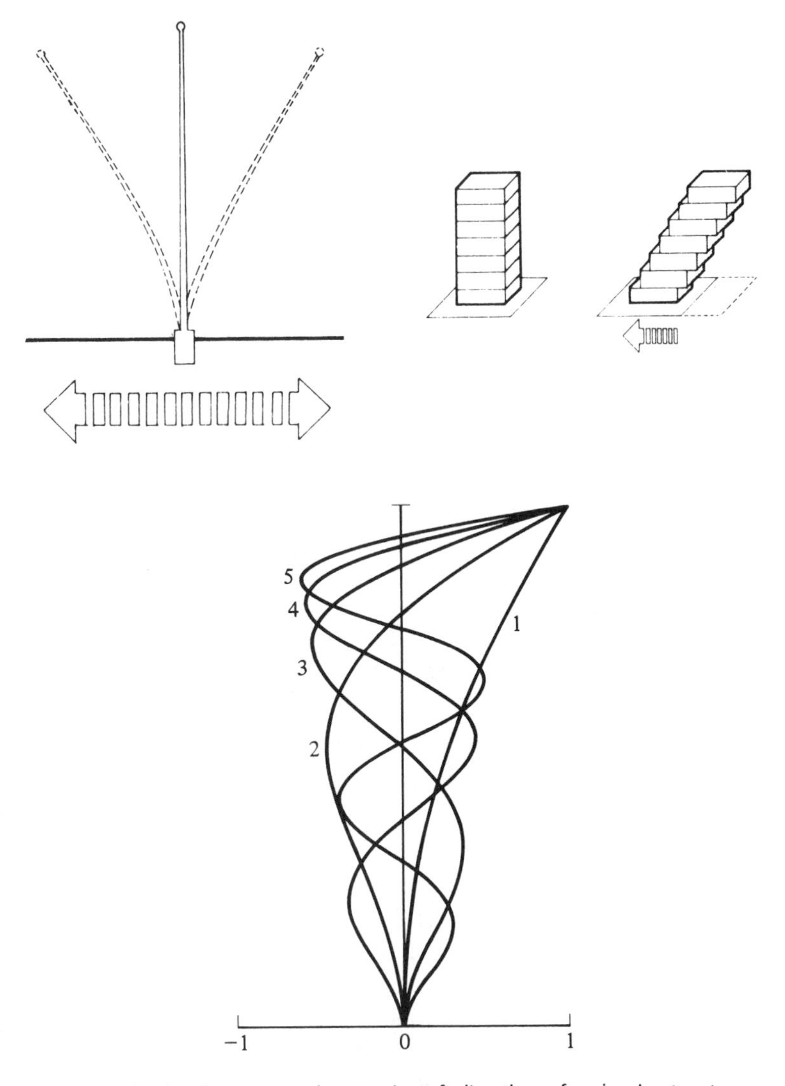

Figure 2-10 Representative modes of vibration of a simple structure.

pending on its configuration. Diverse configurations will have diverse sets of natural modes of vibration (see Figure 2-10 for diagrams of representative modes of vibrations).

When the natural period of vibration of any body (or building system) coincides with the natural period of rhythmic impulses applied by a dynamic force (or earthquake ground shaking), a synchronized resonance between the two results. When this happens, it may be very detrimental to the structure absorbing the dynamic forces, since each successive impulse cycle will cause an increase in the total deformation experienced by a building to a point where the structure may actually be "pushed" beyond its elastic range into greater and greater permanent deformation. The architect must be aware that structural systems and their components may fracture at this point, even before the maximum seismic energy impact is realized, thus precipitating a potential failure or collapse of a building.

A dramatic example of this type of dynamic failure is the famous collapse of the Tacoma Narrows Bridge in Washington in the 1940s, the result of the span going into one of its natural modes of vibrations in resonance with the extreme winds that were buffeting the bridge that morning for some length of time. The bridge was finally pushed far beyond its elastic limit by the continuous rhyth-

Figure 2-11 Example of building collapse due to resonance with long-period ground motions, Mexico City, 1985 earthquake.

mic wind gusts, and total collapse occurred as successive deformations exceeded the rupture point. Similarly, tall structures may go into resonance with long-period earthquake-induced ground motions in the foundation soils underneath and result in total collapse, as was experienced in 1985 in Mexico City (see Figure 2-11.) The architect must realize the importance of this factor in the design of tall buildings. The design of an overall structural system that, during an earthquake, would have the same rate of oscillation, or natural vibration mode, as the "natural period" of the soils on the site should be avoided where the potential of resonance is high.

Building Configuration

Directly related to the negative influence that a structure's natural period/resonance potential may have on seismic performance is the overall configuration of a building's design. Still another design constraint, an understanding of the effect that extreme variations in building configuration or eccentricities in plan shape and vertical mass have on seismic performance, is most important.

By now, all architects know that it is next to impossible to satisfy all outside constraints placed on the design of a building. Design procedures in themselves involve a complex decision-making process filled with variable options, all to be considered and assessed by optimization analysis. Yet, when involved in the design of a building system to be located in a region of high seismic risk, it is quite clear that the life safety aspects of a structure are to be ranked as a high-priority item. Consequently, the impact of building configuration on seismic performance becomes a fundamental constraint placed on the architect during the initial design phase.

During the preliminary design of a building, the overall layout and basic masses of the structure are determined in answer to the client's programmatic requirements. All design decisions made at this time profoundly effect seismic performance. Given that earthquake ground waves may arrive at the site from any direction, the building system must be able to oppose lateral loads coming in from any point. Obviously, the best approach to this type of a problem is to design a structure able to withstand forces from any and all sources.

The implication for architectural design is that the building system best suited to solve this problem is one that is symmetrical about both axes in plan and without irregularities or setbacks in elevation (see Figure 2-12). Constraints imposed on the design of a building by site conditions such as shape and topography, the client's requirements, local ordinances, and programmatic criteria constantly interfere with achieving such a formal design approach demanded by symmetry in all directions. From the very beginning of the design process, therefore, a comprehensive awareness of

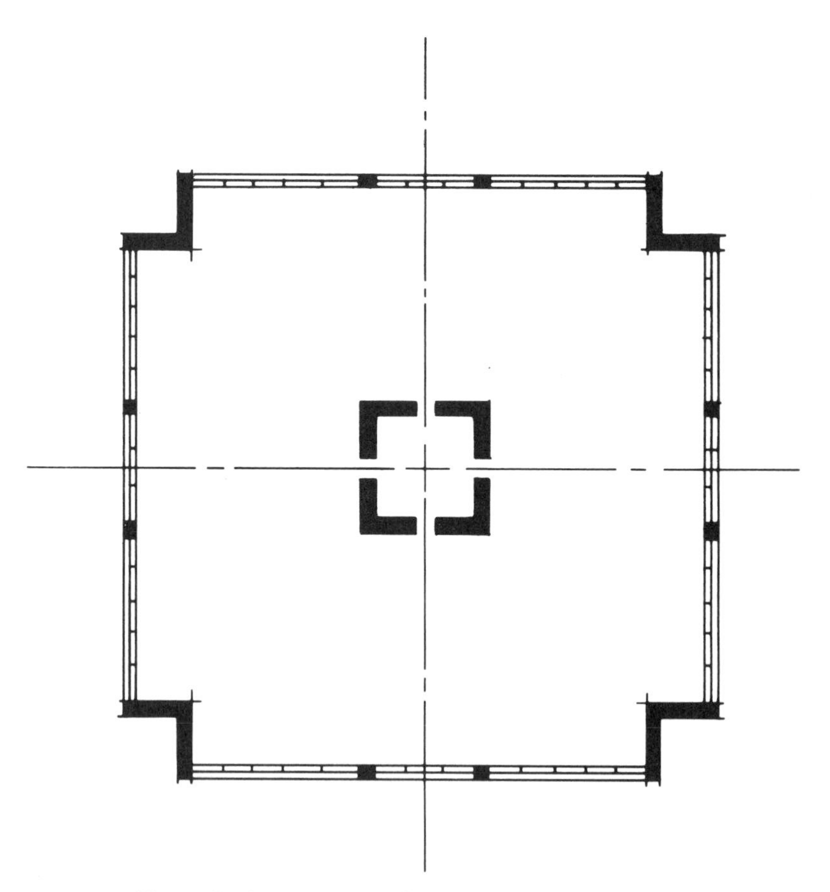

Figure 2-12 Floor plan of a symmetrical building layout.

how building variations influence the basic seismic performance of a structure is absolutely necessary. From this point of view, it is strongly recommended that the architect work in tandem with the consulting structural engineer on the building system's design from the inception of the design process. By working closely together when initial decisions are being made and while many design options are still available, counterproductive design strategies may be avoided.

Many field studies conducted after an earthquake indicate the adverse effects that configuration irregularities in building plan arrangement and evaluation layout have on seismic performance. Such building irregularities are often responsible for major damage during a severe earthquake, as they may lead to detrimental torsional effects that were not considered during the design process. Torsion occurs in a building subjected to dynamic loads when the center of mass does not coincide with the center of rigidity, causing the potential for rotation to occur about its center of rigidity. It has been observed that this often leads to increasing the effect of lateral forces on structural components in direct proportion to their distances from the center of rotation.

Three examples are used to illustrate this problem. The first assesses the seismic performance of an "L"-shaped building. Each leg of the L shape will experience diverse deformation demands in direct correlation to its respective position relative to the incoming direction of the earthquake loads (see Figure 2-13). Under the influence of the earthquake force coming in from the direction shown, Wing A, parallel to the direction of the force, will be stiffer because of its more rigid and resistant-prone axis than Wing B, perpendicular to the earthquake motions, which is shallower in that direction and therefore feasibly more flexible and/or weaker in its seismic performance. In addition, this type of plan configuration sets up a situation in which undesirable torsional forces are introduced into the building system under the influence of earthquake motions, causing rotation of the mass of Wing B relative to the center of rigidity established by the L shape. Unless the two wings are designed with the capacity to resist and dissipate these torsional effects adequately, the building system may absorb severe damage, particularly at the notch where the wings meet. Figure 2-14 shows the result of the seismic performance of an L-shaped building, after the 1971 San Fernando earthquake, in which one wing survived ground shaking while the other wing collapsed.

The second example takes up the case of two multistory bank buildings located in the city of Managua, Nicaragua, which was severely shaken by the major, damaging 1972 earthquake. Figure 2-15 shows the floor plan layout of the Banco Central. The second building, Banca d'America, had a square-shaped, symmetrically laid-out plan with a stiff elevator core placed in the center (see Figure 2-12). In contrast, the Banco Central, although a regularly shaped building, was rectangular in plan layout, with the stiff elevator core located at one end. The two sites had similar geophysical characteristics, as the two buildings were located within five blocks of one another, so soil failure was not a consideration. The building type of the two buildings was similar, and type of construction or use of building materials was not a factor, since both were well-designed frame buildings.

The Banca d'America absorbed little damage, was repaired after the earthquake, and is still in use today. The Banco Central suffered

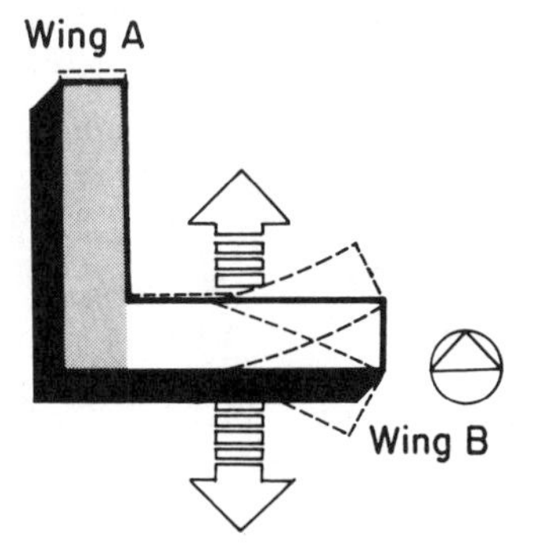

Figure 2-13 L-shaped floor plan configuration. *Source:* Botsai, et al. (1976). Reprinted with permission.

Figure 2-14 *Top:* Relatively undamaged wing of an L-shaped building. *Bottom:* Other wing of L-shaped building that collapsed.

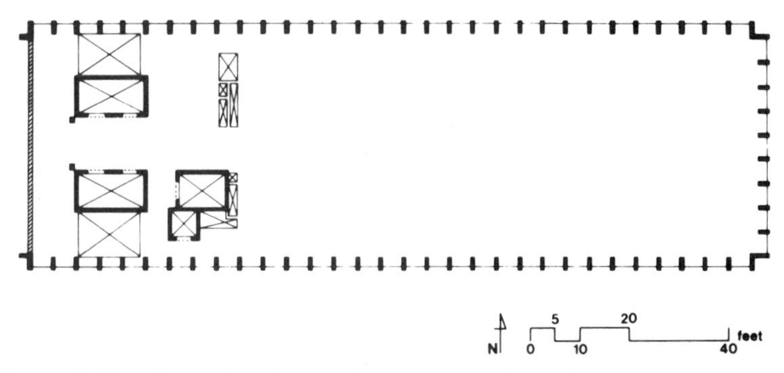

Figure 2-15 Floor plan shape of Banco Central Building, Managua, Nicaragua, 1972. *Source:* Botsai, et al. (1976). Reprinted with permission.

major damage that was not considered repairable and was subsequently demolished. A thorough analysis completed after the earthquake indicated that Banco Central developed unrestrained torsional forces as the free end, more flexible portion of the plan shape rotated around the very stiff, off-center elevator core. Owing to its symmetrical plan in both axes, and the centered core, the Banca d'America did not develop torsional forces and survived with minor damage. The results of the seismic performance due to configuration in plan shape speak for themselves.

The third and final case study presents two building examples dealing with variations in elevation. Figure 2-16 offers diagrams of

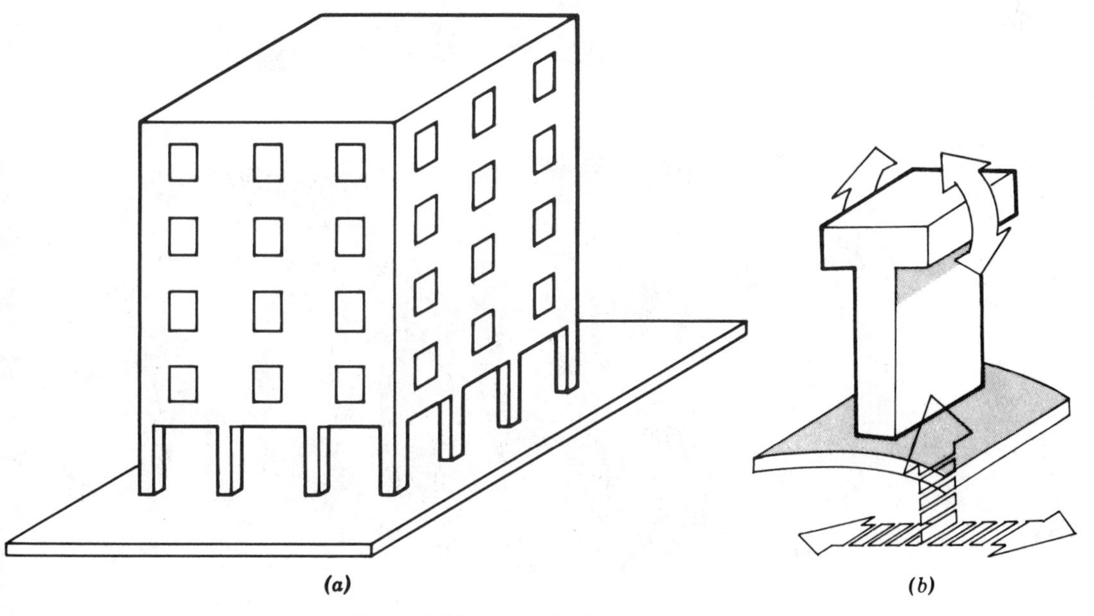

Figure 2-16 *(a)* A "soft story" building; *(b)* reverse pendulum building.

two buildings with discontinuities in their vertical projections: (1) a flexible, "soft-story" building; and (2) a reverse pendulum building. Both represent potential problems in dealing with the effects of building configuration and seismic performance.

The first, a "soft-story" building, is defined as a structure with a stiff, rigid superstructure placed on top of an open, flexible first floor with relatively unbraced vertical supports between large openings. Structural discontinuity occurs between the open first floor and the stiffer upper floors, as the loads carried by the superstructure down to the lower floor of the building must make an adjustment and be channeled through a stress concentration at the transition point. Unless the transition point has been adequately designed to absorb these stress concentrations and/or allow for the transition of forces to the vertical supports at the lower level, failure will occur (see Figure 2-17). This also is exactly what happened to several four-story wood-frame condominium buildings in the San Francisco Marina District during the 1989 Loma Prieta earthquake when the soft first story of the building, punctured on

Figure 2-17 Olive View Hospital, 1972 San Fernando, California, earthquake.

two sides by numerous garage doors, failed under the loads superimposed by the three-story stiff, box wall system above it (see Figure 12-13).

There are many multistory buildings of this soft-story type located in our metropolitan centers, since it remains a popular architectural solution to programmatic requirements faced by the architect, one of which often is to design the first story with as much openness as possible in order to "attract and draw" the pedestrian or passing motorist into the interior. Automobile showrooms, department stores, (with their display spaces), and commercial exhibition centers are typical of structures that require an exterior treatment of this type on the ground floor. And now it is not uncommon to find such soft-story buildings designed as part of a multistory mixed-use complex, with corporate office spaces or condominium units above.

Two appropriate solutions, among others, to resolve this soft-story deficiency would be to eliminate: (1) the discontinuity by carrying the same stiff, box shear-wall system down through the first story, or (2) eliminating stress concentrations accumulating at the transition point by increasing the size of the columns and thus easing the flow of forces around that critical area.

The seismic provisions of the 1988 Uniform Building Code (UBC) recognize the importance of building configuration on performance levels. In various tables in UBC Section 23, the complexities of building configuration are defined, and specific measures are given to indicate when offsets torsional irregularities, reentrant corners, soft stories, and diaphragm discontinuities become critical (see Figure 2-18). These examples, as illustrated in the accompanying figures, should be sufficient for the architect to understand the basic principles involved in the impact of building configuration on the general, overall seismic performance of a building system. For those who wish additional information on the subject, see Arnold and Reitherman (1982).

Relative Rigidity

In any structure, earthquake loads will automatically focus on the stiffer, rigid elements of the building system. The less rigid, more flexible building elements have a greater capacity to absorb several cycles of ground motion before failure, in contrast to the stiff, more brittle elements, which may fail abruptly and suddenly shatter during an earthquake. Everything else being equal, in a structural combination composed of both rigid and flexible elements, the less rigid elements will tend to pass the seismic loads on to the more rigid building components, whether they are designed to resist the loads or not. When this accommodation occurs, a concentration of stresses is again found to focus on the more rigid elements, resulting in their failure owing to their inability to resist high stresses for

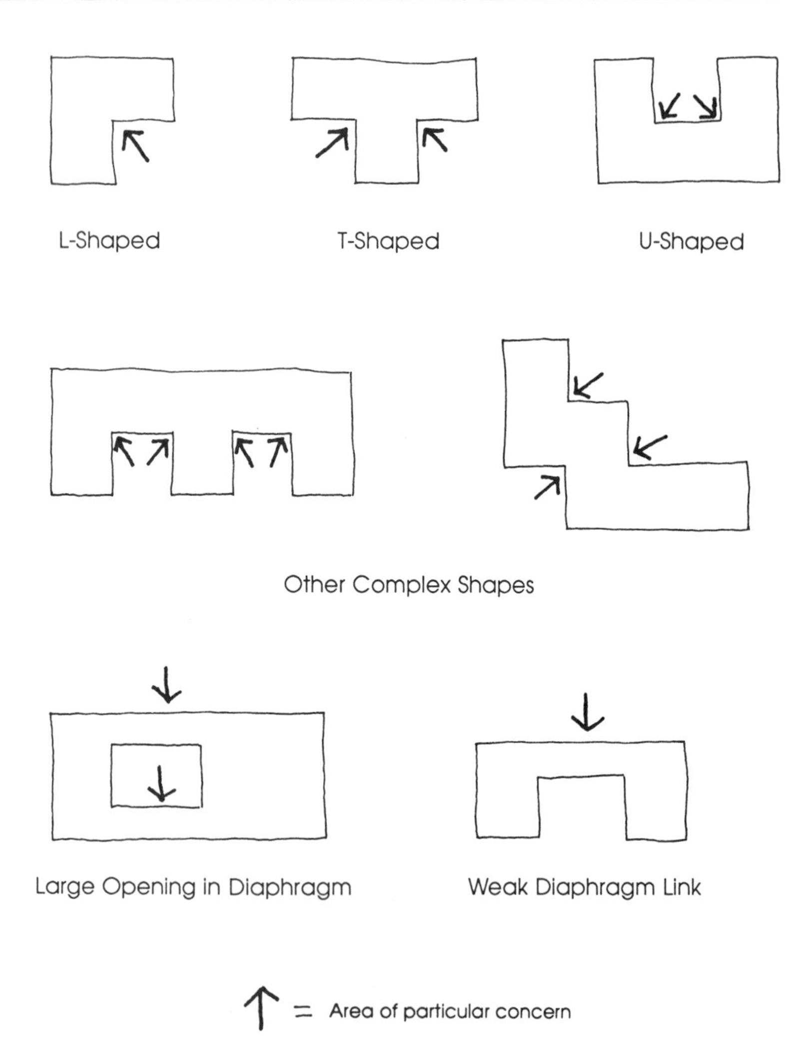

Figure 2-18 Diagrams of irregular floor plan configurations and floor plan shapes with reentrant corner concerns. *Source:* Rojahn and Reitherman, (1989), ATC-20, Applied Technology Council.

which they were not designed. Thus, the relative rigidity of building components becomes a very important topic in seismic design.

In postearthquake field inspections, many examples of this type of failure are seen. One of the most common examples is related to the "short-column" effect. It is dramatically illustrated by the failure of a reinforced concrete-frame building used as a two-story parking structure in the city of Whittier, California. Owing to accommodations made for a sloping site, the vertical supports at the first story of the parking structure were designed as a series of columns that were long at the high end of the building and short at the low end of the ground floor, which followed the slope of the natural grade. The roof level was designed as a flat horizontal plane for ease of parking. This design solution set up a perfect illustration of the short-column effect.

When the 1987 Whittier-Narrows earthquake struck, the longer, more flexible columns at the higher front of the building indiscriminately passed on the lateral loads to the stiffer short columns in the back instead of distributing the loads equally among all of the columns. Automatically, the short columns were overwhelmed by a concentration of stresses for which they were not designed. Failure occcured along the line of short columns, as shown in Figure 2-19. The longer, more flexible columns simply deflected without cracking, since they were not overloaded and passed their portion of the lateral loads on to the stiffer columns. Soon after the earthquake, after being declared a total loss and a hazard to life safety, the building was demolished as quickly as possible.

In other circumstances it has been determined that nonstructural components of a structure have been required to carry unintended lateral loads simply because they were the most rigid elements in the building system. In one multistory building severely damaged by the 1964 Great Alaska earthquake, short, stiff concrete spandrals that were intended to be nonstructural were placed between reinforced concrete columns. The flexibility of the less-restrained taller columns simply transferred the loads, which crushed the spandrel panels in a typical shear-cracking damage pattern (see Figure 2-20).

Another example of this nonstructural complicity is found in the performance of rigid exterior infill panels placed within a multi-

Figure 2-19 Damaged parking structure after the 1987 Whittier-Narrows, California, earthquake. *Source:* Paul Neel.

Figure 2-20 Typical X-shear cracks in building spandrels after 1964 Anchorage Alaska, earthquake. *Source:* K. V. Steinbrugge, private collection. Reprinted with permission.

story frame system. When the panels have been rigidly attached to the structural frame without any precautions taken to protect their integrity, movement of the more flexible frame has in some instances caused infill masonry panels to be crushed by the frame action and literally explode outwardly into the ground below. During the 1985 Mexico earthquake, many buildings suffered extensive damage of this type, resulting in considerable heavy debris cascading into the streets and walkways below.

Typical solutions for mitigating this type of damage include either: (1) designing the infill wall panels to be free from frame action by providing sufficient space between the columns, upper floor, and walls to avoid pounding damage to the columns and/or panels; or (2) providing a containment to the wall panels so that they won't explode when stressed by the movements of the frame attached to the structural system.

However, there is another side of the coin that must be mentioned as regards this problem. The desirable effects of energy absorption and dissipation by a building system under earthquake loads, mentioned earlier as a positive approach to seismic design, rely on all elements of the building, structural and nonstructural, assuming an active role in a damping effect against earthquake forces. If nonstructural components are heavily damaged during an earthquake but do their part in absorbing energy and damping the effects of the earthquake in preventing total collapse and severe life loss, it is argued that the seismic design concept has been successful. However, it is acknowledged that there is a fine line of distinction to be drawn here; the concern is that the concept will become so sophisticated that damaging surprises will happen unexpectedly and catastrophically. Redundancy in a building system is an asset when undertaking the seismic design of structures of all types.

Damage Control Considerations

In the distant past, the general philosophy governing the seismic design of buildings has been clearly focused on public health and safety. The life safety aspects of building design were acknowledged to be of the highest priority. In the mitigation of earthquake hazards, the threat of life loss has of course been identified as the most important factor, one that has led to the successful promulgation of initial seismic safety performance standards since 1933. It was the moving force behind the National Earthquake Hazards Reduction Act, a national program that Congress passed in 1976. Even in 1933 it had been the rationale for passage of the Field Act and the Riley Act, which proposed higher performance standards as a required component in the seismic design of public schools and public buildings in the state of California. Yet, passage of these two acts also established the precedent that it was perfectly and publicly acceptable to require higher standards for the design of certain building types and not for others.

Currently we find ourselves embracing this philosophy in the design of structures housing that are now defined as critical emergency service facilities. In 1972, after the disastrous performance of medical service buildings during the 1971 San Fernando earthquake, the California state legislature passed the Hospital Act, which required that all public general hospitals and other critical medical facilities be designed at a higher seismic performance standard than that of other buildings. The highlight of the act was that all major hospitals were to be designed to a performance level that would ensure their operation after a severe, damaging earthquake. A clear and distinct emphasis was placed on the words "continued operation" as a guarantee that hospitals would continue to perform all their functions after such an event. Thus, for the first time "damage control" was introduced as a significant factor in the design of buildings. The distinction was not arguable, nor in a similar fashion is the seismic design of nuclear power plants or other critical emergency service facilities. Such higher design standards are publicly accepted in good faith.

The philosophy of "damage control" in seismic design is currently becoming of greater importance. The principal reason for this is that the complexity of our buildings is ever-increasing as electronic communications systems and the accompanying hardware become a necessity for any business or function. One need

Figure 2-21 Tilted building after 1967 Caracas, Venezuela, earthquake. *Source:* U.S. Geological Survey and NOAA/EDIS.

only look at the large amount of electronic data-processing equipment, and its high investment cost, now considered essential in any new office building. We now have a situation where damage to the building itself must be controlled so that cascading debris does not wipe out the investment in equipment and communications systems. This subject will surface again in Chapter 6, where the nonstructural components of a building will be examined in detail.

Site Conditions and Location

The final issue to be addressed as a factor influencing building performance deals with site planning considerations. The topic will be touched on briefly here, as the next chapter will cover the material more fully. Suffice it to say here that the best building design approaches, even those using the highest seismic standards available, are useless unless attention has first been paid to the location and site planning of the facility.

There are many examples wherein outstanding buildings, well constructed and adequately design to resist seismic forces but planned without any consideration given to the geological characteristics of the site, have been rendered useless and inoperable after an earthquake. Figure 2-21 illustrates an example of a well-designed building rendered uninhabitable after the liquefaction of soils on the property. Earthquake-induced landslides, local soil failures such as liquefaction or subsidence, and flooding can be significant threats to the serviceability of any building.

Prior to starting any preliminary design drawings, a careful as-

Figure 2-22 Broken utility waste line to an office building, Mexico City, 1985 earthquake.

sessment must be made of the site. The integrity of the building system itself depends on the capacity of its foundations to support any loads placed on it by the superstructure above. In turn, if the underlaying soils themselves are incapable of supporting the building's foundations adequately, the building above may become functionally useless even though its well-designed basic structure may still be intact. In several instances where the failure of site conditions led to severe problems, the building became useless when it became inaccessible after sinking into the ground. In other cases, site failures led to the rupture of lifelines "feeding" the building, which became uninhabitable when disconnected from needed utility and communication hookups (see Figure 2-22). And where natural gas lines have been ruptured, as occurred in the Marina District of San Francisco after the 1989 Loma Prieta earthquakes, fire may destroy a building that originally survived the earthquake intact (see Figure 12-11).

CHAPTER

3 SITE INVESTIGATION

As indicated in the previous chapter, site location is just as crucial as building design in the expected seismic performance of a building. Misjudgments in the selection of a seismically desirable location for construction may be as damaging as a fundamental design error in rendering a building inoperable and useless after an earthquake. Accordingly, a comprehensive investigation of the site, even if it has already been selected, is always paramount before starting any preliminary design drawings.

Ideally, the architect should be called in to assist the client in the selection process of an appropriate site as part of his professional services. The client should be aware that a site may dictate many design determinants by placing limitations on the building's size, accessibilitiy, and configuration. Above all else, it certainly will influence the size and characteristics of the lateral forces to be transferred into the building through the foundation system.

SITE CONDITIONS

Local geology and soil conditions of the site will determine the characteristics of earthquake ground motions that may be experienced. Based on the investigation of the local geology and composition of soils in the area, it is possible to anticipate what type of motions will occur at the site in terms of frequency, acceleration, velocity, and amplitude for given earthquakes. For complex sites it is imperative that a professional geotechnical engineer be used as a technical consultant to assist the architect in advising the client on the limitations of the site, to help ensure that any negative aspects are thoroughly understood.

Local soil conditions at a specific site also have a significant effect on ground-shaking amplitudes. Basic rock motions have certain characteristics associated with sharp, high-frequency accelerations and velocity movements. As seismic waves travel through less dense soils, motions are modified by the depth of soil overburden, which increases the amplitude of motion and emphasizes the longer, dominant periods of vibration.

Studies of the 1906 San Francisco earthquake, the 1989 Loma Prieta earthquake in the Santa Cruz Mountains, and the 1964 Great Alaska earthquake indicated that the most damaging motions occur in zones of deeper, less consolidated soils at water-saturated bay margins and marshy areas, in contrast to bedrock sites (see Figure 3-1). In several earthquakes, pockets of severe building damage have been identified in areas where deep alluvium soils were located over bedrock. The effects of deep alluvium soils were in evidence in the 1906 San Francisco earthquake, the 1967 Caracas, Venezuela, earthquake, and the 1985 Mexico earthquake, among others.

Short, rigid buildings may respond negatively to the higher-frequency motions, while high-rise buildings may not be so adversely effected. On the other hand, as observed in the 1985 Mexico earthquake, low, rigid buildings were not critically affected by the long-period ground motions experienced in the old lakebed area, whereas high-rise buildings located in the older landfill areas were subjected to severe damage and/or total collapse.

For the reasons cited, the architect must be familiar with local geological and soil characteristics of the site to know how they will influence earthquake ground motions which in turn will affect building performance. Designing a building system that is not compatible with the frequency of earthquake ground vibrations influenced by soil conditions may clearly cause the building to "tune in" in sympathy with ground shaking and experience larger movements than those for which it was designed.

Figure 3-1 Marina District, San Francisco, 1989 Loma Prieta earthquake.

DISTANCE FROM FAULT

Immediately after an earthquake there is a tendency by the mass media and the public to identify areas closest to the fault as the zones of greatest damage. In their terms, the greater the distance from the causative fault, the less damage there will be. In certain cases, this is a false assumption.

While such popular thinking may hold true for vertical fault planes, where experience has shown that damage for a few miles on either side of the fault is more or less constant, it is not true for thrust faults, which are characterized by a dip angle, as shown in Figure 3-2. This was the situation experienced in the 1971 San Fernando earthquake, where some zones of heavy damage were miles away from the thrust fault and/or epicentral region. Consequently, clear distinctions must be made between the fault types located in the general region. In addition, the architect must remember that in the West Coast of the United States, southern Italy, Greece, and other highly active seismic areas, the farther the building site is located from one fault zone, the closer it will be to another. In any event, the greatest amount of damage, by far, is caused by ground shaking, which covers a much larger area (see Figure 3-3).

The rule of thumb and bottom line in all cases, therefore, is not to build directly over a fault trace. Although the number of potential buildings damaged from surface faulting may be low, damage to any building located directly over a fault trace will be most severe during a major earhtquake where surface rupture occurs.

LIQUEFACTION, LANDSLIDES, AND SUBSIDENCE

As urban development continues, we find ourselves pushing out into areas that have less desirable building sites. Areas once thought marginal for building construction are now being considered for large-scale development. While it is true that great improvements in construction technology and design methods have allowed us to build on sites rejected in the past, the dynamic motions produced by earthquakes are not sympathetic to poor soils areas.

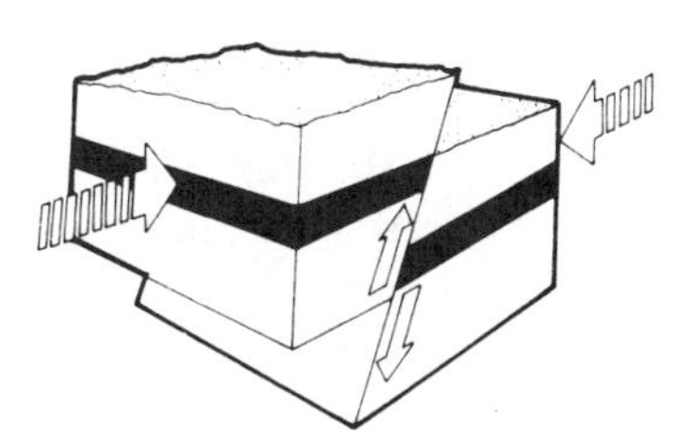

Figure 3-2 Thrust fault Characteristics. *Source:* Botsai, et al. (1976). Reprinted with permission.

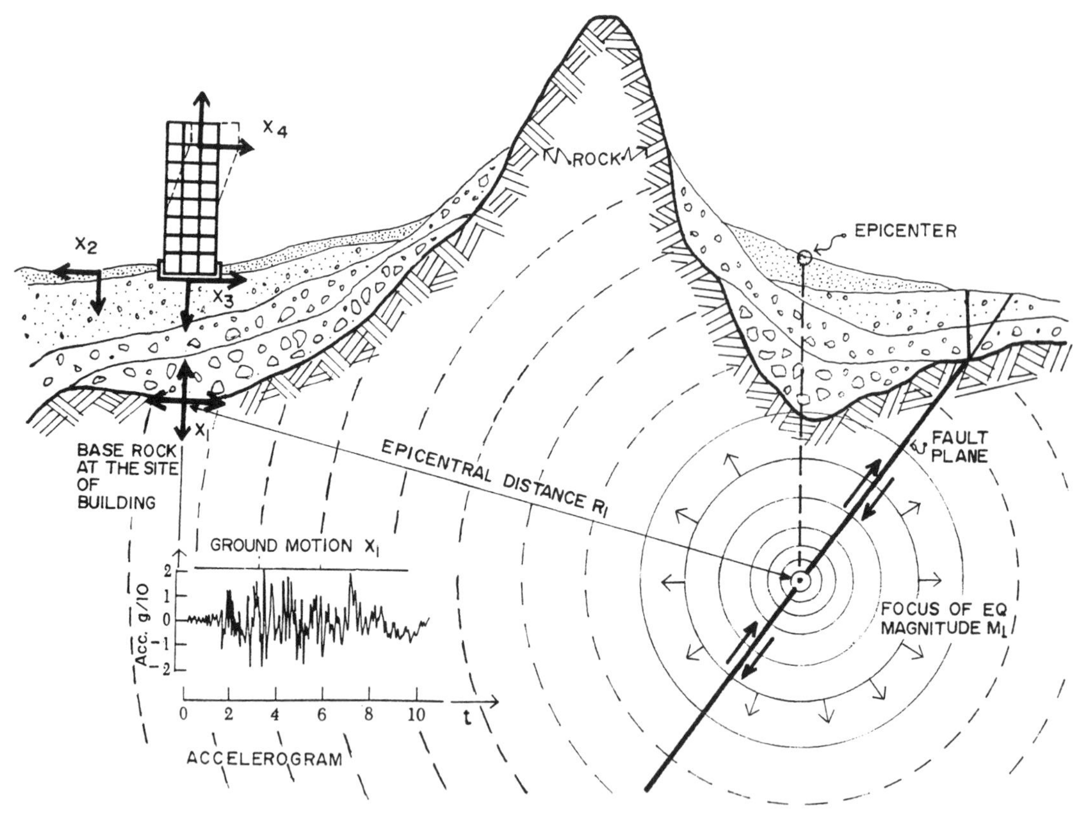

Figure 3-3 Fault relationship to ground motions and building response. *Source:* Guevara (1989). Reprinted with permission.

Liquefaction

The phenomenon known as liquefaction is related to poor soils areas composed of water-saturated fine sands or silts. Under simple, normal static loading, these areas have a reasonable capacity to support a normal building as long as they remain stable. Liquefaction of soils occurs when a material of solid consistency is transformed into a liquefied state as a result of increased pore water pressure in between the fine silts or sands. Water-saturated, granular sediments such as silts, sands, and gravels, which are free of clay particles, are more susceptible to liquefaction, especially the looser, finer materials described.

Although areas of potential liquefaction may appear sufficient to carry loads, in reality their load-carrying ability is deceptive. Under dynamic shaking as experienced during an earthquake, their soils may be rearranged and in doing so lose the capacity to support the foundation systems above them. In effect, such water-saturated soils become liquefied when shaken, and behave as a dense fluid rather than a solid mass with support capabilities. When this oc-

Figure 3-4 Earthquake-induced liquefaction, 1964 Niigata, Japan, earthquake. *Source:* NOAA/EDIS.

curs, buildings may sink into the ground, as their foundations are no longer supported by the soils beneath the structural system. Many examples of this phenomenon exist, the most dramatic illustrated by buildings rendered inoperable during the 1964 Niigata, Japan, earthquake and the 1964 Great Alaska earthquake (see Figures 3-4 and 3-5).

Typically, the liquefaction potential of soils increases with the amount of water found in the layers of fine silts. These types of soils are commonly found in the historic margins of ocean bays or inlets. It is often difficult to identify these areas because, over the years, they may have been filled in and extensively developed so that the external appearance has changed. In the San Francisco Bay Area, two metropolitan airports have been constructed on landfill placed over old mud areas highly susceptible to liquefaction. During the 7.1 magnitude earthquake in October 1989, there was enough damage to airport runways and control towers that both airports' normal operations were shut down for several hours. One of the most dramatic examples of damage due to liquefaction was seen during the 1960 Chile earthquake, when underground liquid storage tanks, which were empty or partially empty at the time of the earthquake, rose to the surface of the ground when the soil in which they were buried liquefied under seismic ground motions.

Landslides

Earthquake-induced landslides are quite common during major events. Again, as in the case of liquefaction, the stability of hillside

Figure 3-5 Turnagain Heights ground failure, 1964 Anchorage, Alaska, earthquake. *Source:* NOAA/EDIS.

areas can be deceptive in that what may appear to be stable under gravity loads may not be stable after all under dynamic ground shaking. This is especially true for hill areas during a very wet, rainy season of the year, when hillsides are more prone to slide even when not under dynamic excitation.

While it is quite common to find residential-scale construction on hillsides, today it is not unusual to find commercial office buildings, regional shopping centers, professional service hubs, or research and development firms sited in hilly areas. Accordingly, we find more and more of the existing building stock being located in developing areas having a greater exposure to landslide potential than in the past. In investigating the suitability of a site for development, it is increasingly important for the architect to assess the potential of landslides occurring on or adjacent to the parcel under consideration. Figures 3-6 and 3-7 are views of common types of earthquake-induced landslides. These are quite frequent in hill areas and are more apt to occur during the wet season of the year. The figures also indicate typical characteristics found in such landslides.

Figure 3-6 Massive landslide in mountains near Senerchia (a hill town), 1980 Campania-Basilicata, Italy, earthquake. *Source:* Mader and Lagorio (1981). Reprinted with permission.

In this book the natural and physical characteristics of landslides are not important, but building damage patterns caused by earthquakes should be of great interest to the architect. There are many examples of earthquake-induced landslides that caused major damage to building facilities. The most dramatic of these are two examples illustrating different effects of damage due to landslide. In the first, which occurred during the 1970 Peru earthquake, a single devastating earthquake-induced landslide that originated on Huascarán Mountain seven miles away destroyed two complete villages, Yungay and Ranrahirca. It is said that the debris from the landslide fell from a height of 12,000 feet and traveled 7 miles at an average speed of approximately 200 miles per hour.

The second example relates to a school building destroyed during the 1964 Great Alaska earthquake. The site of the building was located in the zone of a massive landslide and subsidence triggered by the 8.6 magnitude earthquake. In such a case as this, nothing could have been done in the architectural or structural design of

Figure 3-7 Landslide blocking access road to Cochrane Bridge, 1984 Morgan Hill, California, earthquake. *Source:* U.S. Geological Survey.

Figure 3-8 Government Hill School Building destroyed by massive earthquake-induced subsidence trough, Anchorage, 1964 Alaska earthquake. *Source:* Henry J. Degenkolb.

Figure 3-9 Sinking of multistory office building due to earthquake-induced subsidence, Mexico City, 1985 earthquake.

the building to resist the forces placed on the building system when the ground slid out from under the foundation supports. It was simply a case of bad siting (see Figure 3-8).

Subsidence

Subsidence refers to the gradual settling of the ground over a long period of time. Uniform subsidence over an entire site may not pose a direct threat to the building itself, except for access to the building and/or the shearing of utility service lines to the building, since the building may settle evenly on the site as it adjusts to new ground level (see Figure 3-9). But when it is not uniform, damage to the building may be severe. Earthquakes simply hasten the amount of subsidence that may take place throughout a given site, and may introduce differential settlements that can cause severe damage by racking a building out of shape, cracking floors, walls, ceilings, and roofs. The dynamic ground motions produced by an earthquake simply compound the problem by introducing highly concentrated stresses, at various points of the structure, that the building was not originally designed to resist.

High stress points caused by differential settlement over the years may weaken a structure substantially and be the main cause of major damage during an earthquake. There is no doubt that buildings previously weakened by any cause will not have an impressive performance record during an intense seismic event.

CHAPTER 4

SITE PLANNING

When a client approaches an architect for a proposed building program with the site already selected, among the first priorities to be discussed is whether an assessment of the site, adjacent properties, and the region as a whole has been completed to determine the existing earthquake hazards. This must be done even before the preliminary design studies for the building and/or preliminary planning studies for the site are started.

If a hazards assessment has not been made, one should be done immediately without further discussion. As indicated in the previous chapter, the geologic characteristics of the site will have a direct bearing on the effects an earthquake will have on the total building system. If a hazards assessment has already been completed, then an analysis of the effects, if any, that an earthquake will have on the performance of a proposed building on the site should be made. In either case, it is first necessary to take predesign actions to determine (1) what should be expected at the site if an earthquake occurs during the lifetime of the proposed building, and (2) what the effects would be of such an event on the building's performance.

What is essential at this point in the design process is that a general seismic risk assessment of the site and the region be conducted in order to obtain a detailed understanding of potential exposures to earthquake and other geological hazards. At this stage the assessment is limited to a general one only and not site-specific in terms of core drilling for soil samples and/or trenching. It represents an investigation to answer the question, "What is the relationship of the site to earthquake and geologic hazards locally and regionally that will influence the design and construction of a building, or a complex of buildings, on the property?"

Obviously, without incurring high costs of drilling or trenching, one of the critical things that should be determined is simply whether or not a fault runs through the site. If there is a fault, the next questions should concern its locatation on the site, its characteristics, and whether it is active.

If no active faults run through the actual site, other questions still remain if the site is located in a seismically active region: "What is the distance of the site from the nearest active fault, if any?" and "If there is an active fault in the area, what is the maximum probable event that may occur, and what would be the maximum probable peak acceleration that could result?" We are dealing with very pragmmatic consequences and values here, all of which would have a critical impact on site planning and the building's design, construction, and performance. In effect, we are attempting to determine whether or not it is possible and/or cost-effective to build on the site.

To assist the architect in making this determination, there are many official documents available in the public domain. At this point, most of the documents that are helpful during this initial hazards assessment of the site offer general data rather than site-specific information. Nonetheless, it is important that all data sources be utilized. There is always the chance that this preliminary assessment will indicate that the site has no exposure whatsoever to seismic hazards, thus precluding any need for further investigations such as costly trenching or core drilling from the earthquake-risk point of view.

GEOLOGIC MAPS

Of the many official documents on the topic available to the architect for review, the first that should be examined are the general but technical maps that indicate prevalent geological characteristics on a regional basis. Of these, the two most useful series are those prepared by an agency of the federal government for the entire country and an agency of the state for the entire state. Although quite general, they represent excellent sources for the initial identification and assessment of fault zones and soil characteristics in the area.

On the federal government level, the lead agency in earthquake geologic studies is the United States Geological Survey (USGS). As one of its services, USGS produces a set of topographical maps of the entire country indicating broad geological information, general positions of major earthquake fault zones, and locations of major cities as well as their streets, buildings, transportation systems, and other information drawn at a reasonable scale (see Figure 4-1). In addition, the USGS also issues more detailed seismic maps that specifically identify earthquake fault traces. All such maps are quite helpful in identifying earthquake-prone areas, and will give the architect an immediate picture of the broad, general geologic characteristics and hazards of the region in which his site is located.

The next level of maps are those issued by a state agency. Normally, these maps are produced by a department of geology or nat-

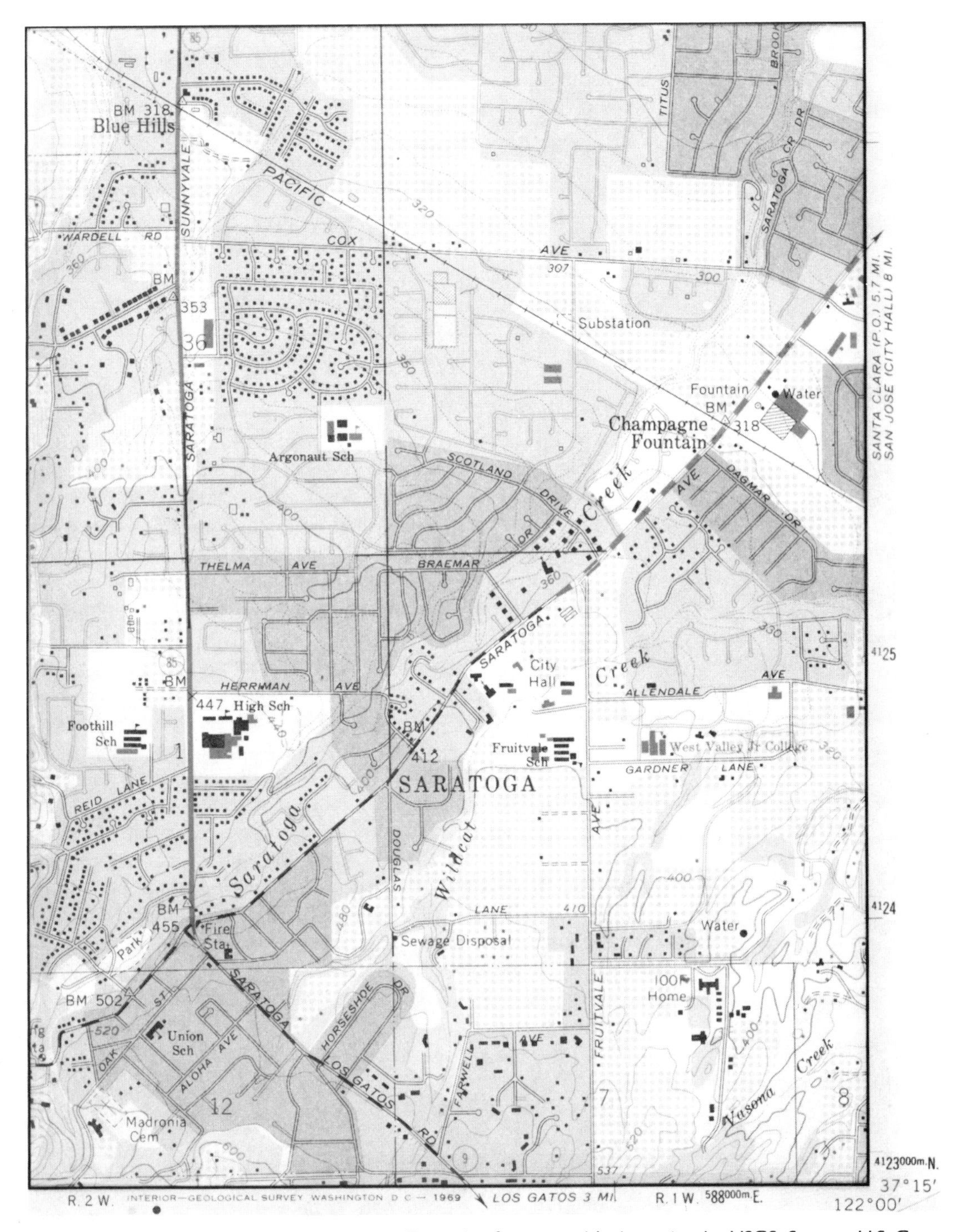

Figure 4-1 Example of topographical map by the USGS. *Source:* U.S. Geological Survey.

ural resources within the state government. For example, in California it is called the Division of Mines and Geology (DMG), whereas in Washington it is called the Department of Mines. In any event, like the USGS maps, maps issued by a state give general geologic information, including the location of fault zones, but in a more specific manner with more detailed information. One such typical publication issued at the state level would be a geological atlas, containing detailed maps of concentrated geological data.

In recent years, there has been a tendency for states to issue maps, as part of land use policy or development goals, for the principal purpose of identifying fault locations; in this way, areas of "special studies zones" have been created. The use of these maps is mandatory for all nonresidential and large-scale residential construction. They indicate in detailed fashion all the potentially and recently active fault lines within the state of California (see Figure 4-2). The operative word in these maps is "active," since only the active fault zones are mapped in detail. The technical definition of an active fault is one that shows evidence of "movement" in recent geological time. Other faults may be undetected or declared inactive, as was the White Wolf Fault that surprised everyone with the 1952 Kern County, California, earthquake. However, even with these limitations, geological maps will be extremely helpful to the architect.

Special studies zones relate to active earthquake fault zones and expressly limit the construction of any building directly over a fault trace. The only building type for which an exception is made is a residential structure up to a fourplex (four-unit dwelling), which is exempted from the restrictions normally applied. If the site is located within the lines of a special studies zone, specific geological site studies are required to determine the exact location of the fault trace on the property. This is accomplished by actual trenching across the site until the exact location of the fault trace is found. At that point, the site plan must take this data into consideration by not allowing any construction directly over the fault trace. Conservatively, it has been suggested that any building on the site be set back an appropriate distance from the fault, preferably in the 50-ft range.

SEISMIC RISK MAPS

Another type of map is one that generally attempts to define the seismic risk of one area compared to another. These maps offer the architect a way of visually and rapidly obtaining a sense of the level of earthquake activity in a broad region. While the USGS geological maps described above offer excellent material on geological characteristics of given areas and the location of fault zones, they are

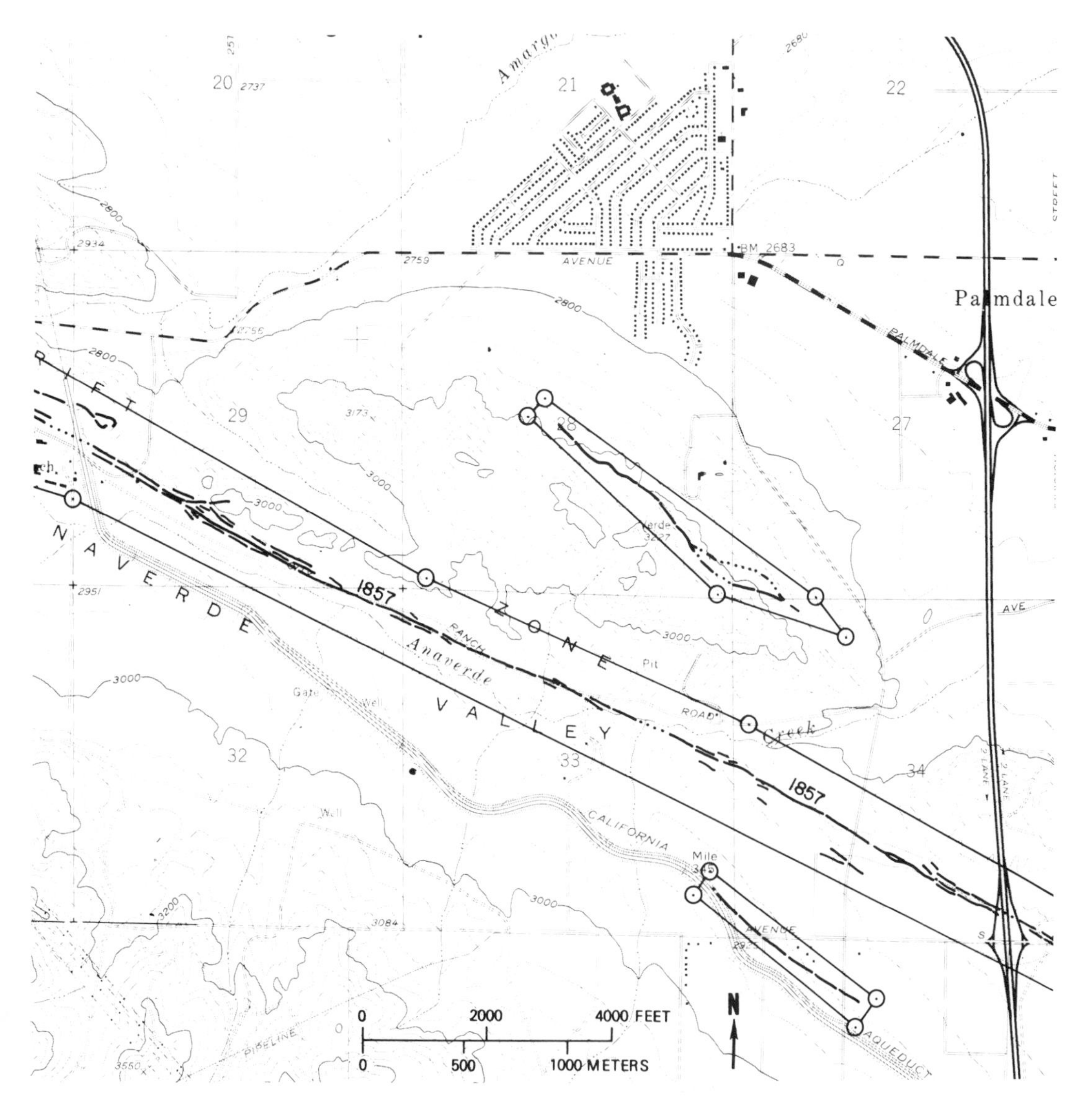

Figure 4-2 Example of mapping of "special studies zones" along San Andreas Fault by CDMG. *Source:* California Division of Mines and Geology.

not intended to give a relative sense of the seismic risk level to which one is exposed.

Generally speaking, the simplest type of national seismic risk map is based on the location, number, and magnitude of historic earthquake events that have taken place and been recorded during about the last 200 years in the United States. These data are then plotted on a map to identify locations having the highest number of earthquakes with the largest magnitudes as those areas at highest risk. The maps are very effective in identifying zones of high seismic risk versus areas of low risk, based on clusters of earthquakes that took place in the past. Such maps are constantly being upgraded on an annual basis, particularly after a year with much seismic activity (see Figure 1).

The data on Figure 1 may then be used to construct a contoured map with numerically designated zones indicating the levels of risk over a large region. In Figure 1-10, which presents the seismic zone map for the entire United States used in the 1988 Uniform Building Code (UBC), there are five zones, labeled 4 to 0. Comparing the maps in Figure 1 with Figure 1-10, the architect should quickly realize that Zone 4 represents the areas with the highest seismic risk.

It has been pointed out that both of the maps described above are somewhat limited for building design purposes because they do not include data giving the expected return period of earthquakes as another measure of seismicity in the region. Nor do they even offer the foggiest idea of what the anticipated peak ground acceleration (ground shaking) might be in an area. Fortunately, in 1978 two new maps were prepared to show the intensity of anticipated ground motions as defined by coefficients for effective peak acceleration and effective peak velocity-related acceleration (see Figures 1-11 and 4-3).

By using all the information offered by these maps, the architect will quickly be able to put together a general regional picture of the geologic characteristics, fault zone locations, level of seismic risk, and effective peak acceleration (ground motion) within a given area. Obviously, this will be most useful in meeting with a client to discuss building design constraints within the general area of the site.

MAPS OF DEFECTIVE-GROUND AREAS

Maps indicating areas of structurally defective ground are another important component to site planning. As a critical factor, many earthquake reconnaissance reports have cited the association between defective ground areas and increased building damage. USGS has published some maps that support this information, but they are developed at a scale that makes their use somewhat limited. However, maps developed by other sources, primarily at the state and local levels, are more appropriate for use by the design professional. There are two types of maps in this category: (1) poor grounds and (2) landslides. Needless to say, when combined with the effects of an earthquake event, both types present areas with potential problems to construction.

In this case, poor grounds are classified as those that are structurally unsound under dynamic loads, such as uncompacted, non-engineered artificial fill areas, and those areas along the original

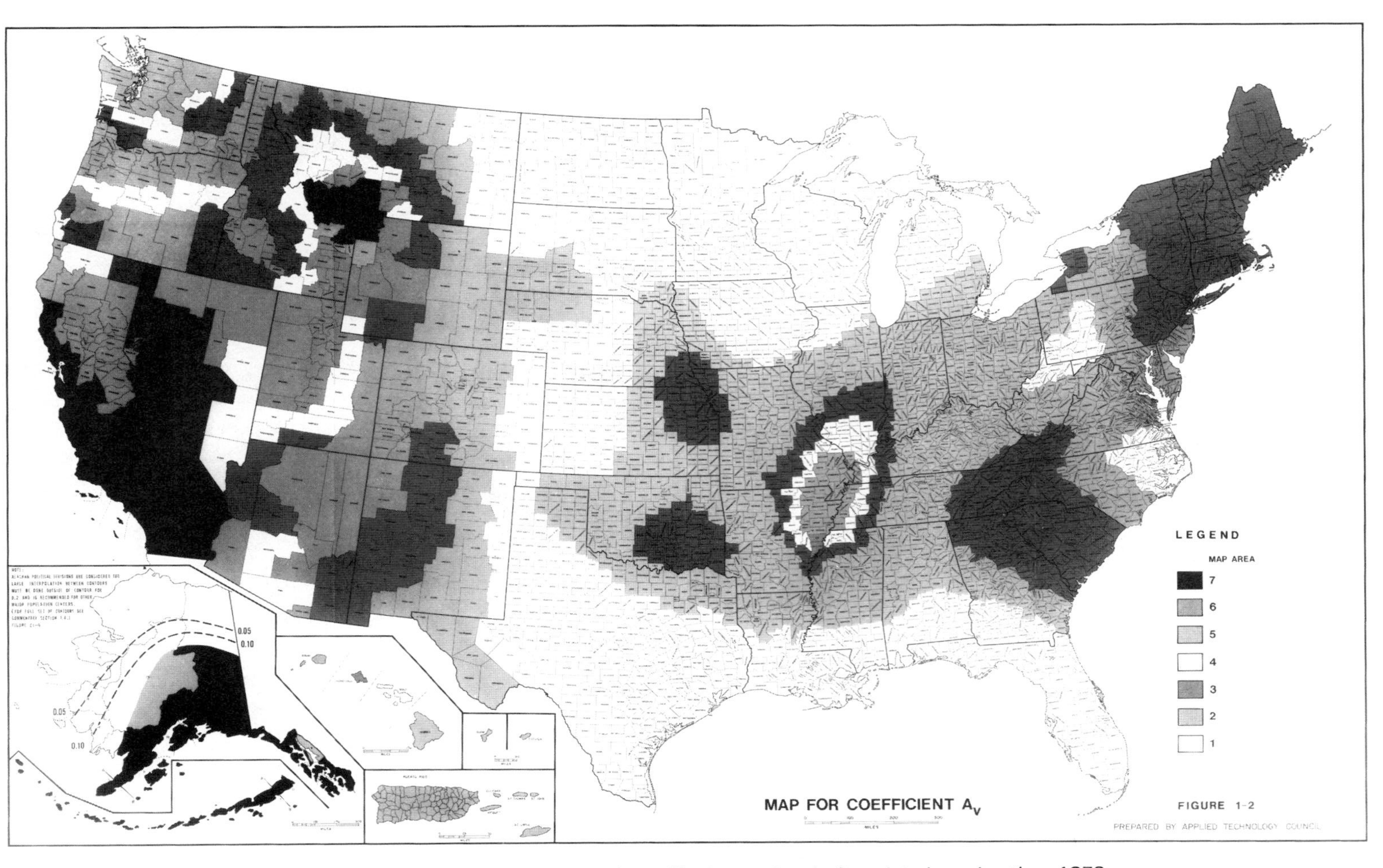

Figure 4-3 U.S. seismic risk map based on effective peak velocity-related acceleration, 1978 ATC. *Source:* Applied Technology Council.

margins of waterways that are water-saturated and susceptible to liquefaction (as described earlier). Ground in these areas is considered to be unconsolidated and unstable, as opposed to firm, undisturbed, and consolidated ground. Typically, they consist of soft, highly compressible, water-saturated fine silts such as those found in marshlands.

Many such areas are discovered around the historic margins of

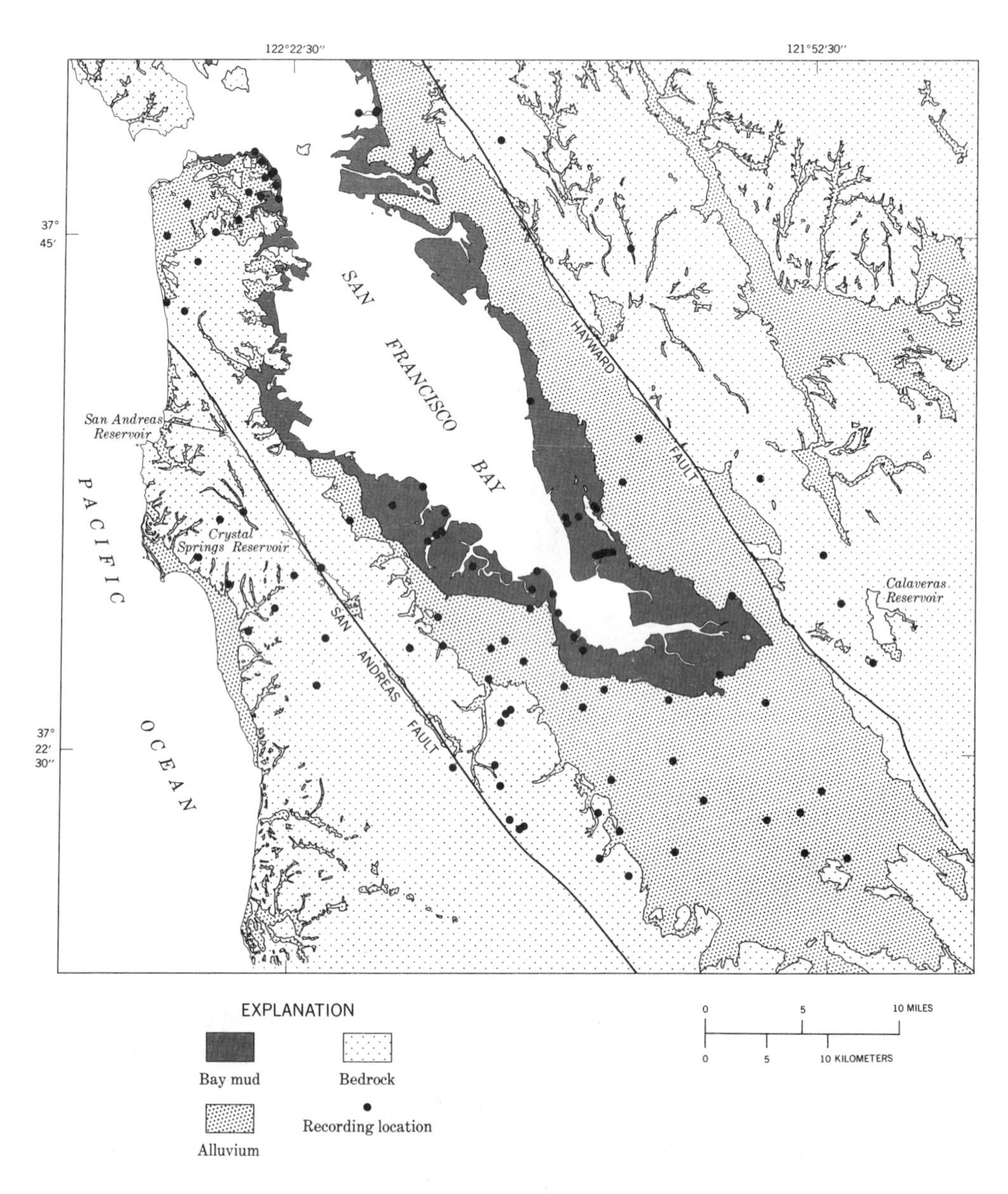

Figure 4-4 Distribution of generalized geologic units, historic bay margins and locations of USGS recording stations, southern San Francisco Bay Area. *Source:* U.S. Geological Survey.

natural bays and harbors that were later reclaimed for use for increased urban development and growth. In many instances the actual demarcation line of the historic bay margin may have been obliterated by urban growth and its exact location forgotten. Coastal cities such as Boston, Seattle, San Diego, San Francisco, Los Angeles, Long Beach, and Charleston (South Carolina), all of which are in seismic zones of considerable risk, are typical examples. An example of a poor grounds map is shown in Figure 4-4. This type of information is very important to the development of a site plan.

Two reasons frequently cited for accentuated building damage in areas of poor soils are that (1) seismic waves transmitted through soft soils are amplified and their period rendered longer, and (2) the building previously may have been weakened due to differential ground settlements prior to the earthquake. In any event, after a severe earthquake it is not unusual to find concentrated pockets of heavy building damage in areas associated with poor soils.

MAPS OF STRUCTURALLY DAMAGING LANDSLIDES

Information on landslide susceptibility that might affect a given area is also available for use by the planning and design professional. Maps of the landslide potential in undeveloped lands are extremely valuable to architects working in concert with developers on large-scale developments, since they indicate the hazard for existing conditions prior to development. Once developed, the hazard will definitely change, either being substantially reduced by appropriate site planning efforts, or increased when basic site planning principles are ignored. As earthquake-induced landslides are quite common, especially during the wet season of the year, it is vital to obtain data on landslide susceptibility before the development of any site plan. This is desirable not only for the building site itself, but also for adjacent properties from which landslides could impact the actual site to be developed.

Typical information on landslide susceptibility is available through the USGS in the form of maps for specific regions in the United States. The U.S. Department of Housing and Urban Development (HUD) also has data available on landslide problems, including a set of maps issued in cooperation with the USGS. One such cooperative effort undertaken in 1972 by the USGS and HUD in California produced two landslide maps under the San Francisco Bay Region Environment and Resources Planning Study. They are

1. Map Showing Distribution and Cost by County of Structurally Damaging Landslides in the San Francisco Bay Region, California, Winter of 1968–1969, by F. A. Taylor and E. E. Brabb, and

2. Estimated Relative Abundance of Landslides in the San Francisco Bay Region, California, by D. H. Radbruch and C. M. Wentworth.

These maps, developed on a general basis rather than a site specific one, nonetheless offer objective information on the landslide-prone areas found in a given region. The readily available information is invaluable in identifying landslide hazards and/or landslide susceptibility prior to the development of a site plan. An example of these types of maps is illustrated in Figure 4-5.

BUILDING ADJACENCY

In dealing with earthquakes, it must be remembered that everything located in the impacted area is oscillating in response to the seismic ground motions and within their own natural period of vibration. This implies that all the buildings in the region will be shaking to one degree or another, not only those on a specific site but also any located on adjacent sites as well. This becomes critical when dealing with high-density, congested urban centers where buildings are often constructed side by side, right to the property lines.

This presents us with a problem, for we now have a situation wherein the performance of existing buildings constructed on the adjacent property lines will have an impact on the new buildings to be constructed in between. As was indicated in Chapter 3, the site planner and architect now have to consider design interrelationships with existing buildings on adjacent properties. This becomes a very important factor in site planning, as it may require a seismic space between buildings, known as a "seismic joint" or "seismic separation," to avoid the potential of severe damage caused by dynamic pounding of adjacent buildings.

To provide an adequate seismic space between buildings, it is necessary to calculate the amount of building drift, or total floor-to-floor deformation, that will occur in the new building and the adjacent existing buildings. The code itself requires a minimum drift limitation for multistory buildings which, as a guide, will give the architect an idea of the amount of space that must be permitted between buildings. It is therefore advisable to keep this concept in mind when developing a site plan for property located in urban centers and in discussing the appropriate structural system with the consulting engineer. This "dynamic pounding" manifestation will be presented in greater detail in chapters that follow.

LAND USE PLANNING AND MICROZONATION

Before turning to recommended principles that govern final decisions to be made on site planning, it is essential to give con-

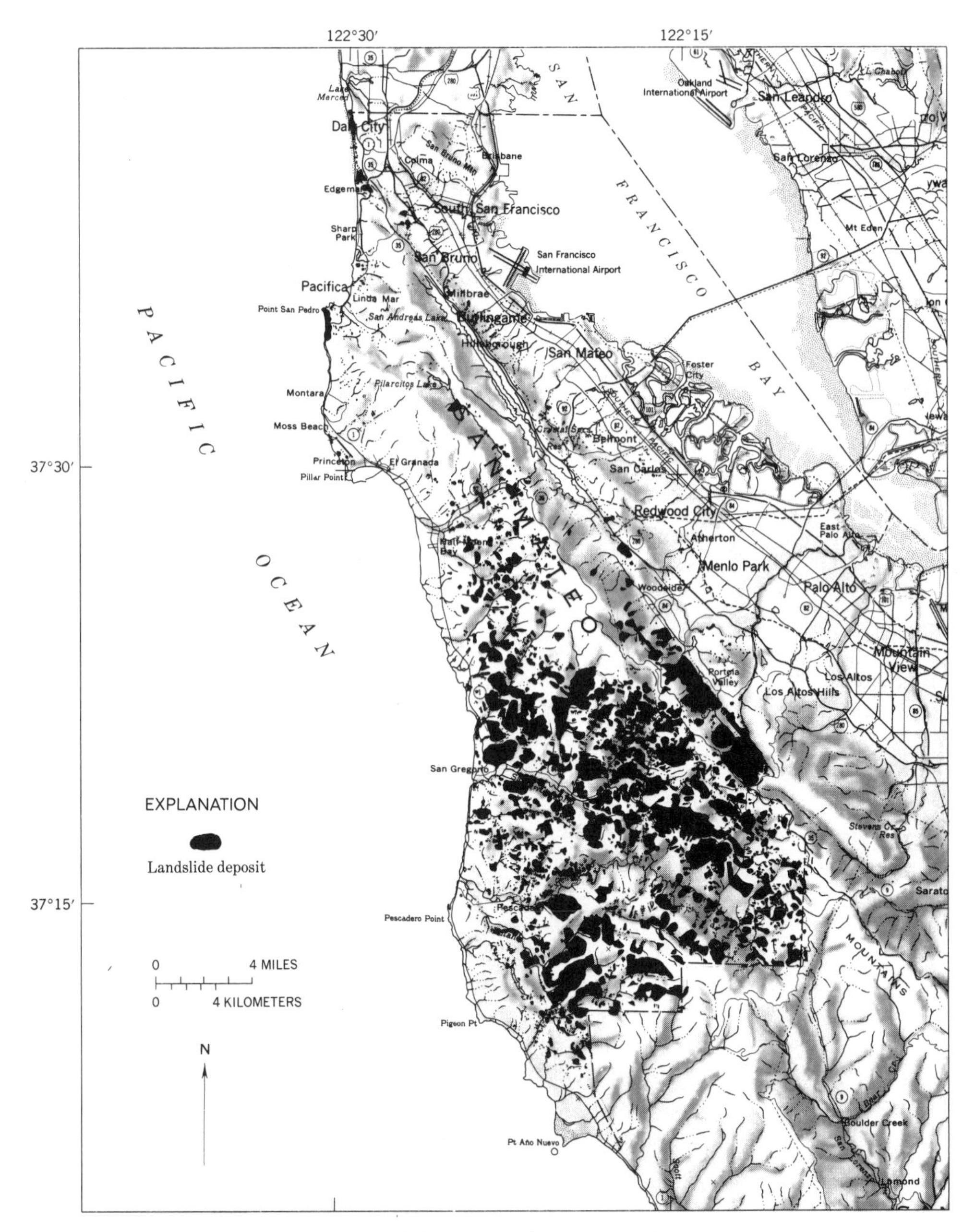

Figure 4-5 Landslide susceptibility map of San Mateo County, San Francisco Bay Area, based on more than 1000 recorded landslides. *Source:* U.S. Geological Survey.

sideration to land use planning policies relative to seismic microzonation. Seismic microzonation methodologies involve the identification, categorization, evaluation, and characterization of all seismic and geological hazards in areas of high or moderate risk, as well as the development of means to coordinate information about seismic risk with land use policy decisions. Normally, such investigations result in maps being constructed from earth-science data collected on a wide scale. These maps typically indicate data on the anticipated maximum earthquake intensity, active faults, geologic units, special studies zones, ground responses, liquefaction susceptibility, landslide susceptibility, and zones of potential tsunami inundation. A good microzonation map delineates areas of potential seismic problems, identifies the problems, and quantifies their effects by indicating their potential severity.

These microzonation maps may then be used for regional land use planning recommendations on the appropriate location of critical emergency facilities, hazardous toxic materials, industrial developments, population centers, transportation routes, communications systems, and open space. The fundamental objective of the maps is the reduction of earthquake hazards through appropriate land use planning on a local level by plotting levels of risk against functional uses. An example of a decision-tree matrix made by the USGS to illustrate this process is shown in Tables 8-3 and 8-4.

SITE PLANNING

Such preplanning studies in seismic microzonation investigations will give the architect a basis for discussions with clients on the suitability of a building site on a regional scale in keeping with the intended building's function and programmatic requirements.

Once the overall perspective of appropriate land uses has been determined, site-specific planning may start if the building site has not been rejected as being incompatible with the intended building occupancy, use, and function. The three steps required in this endeavor are (1) an assessment of a microzonation map in terms of the appropriateness of land use on a regional scale, (2) a site-specific analysis, and (3) an evaluation of the suitability of the specific site. It is important to evaluate the intended building use, occupancy, and function against the seismic constraints imposed on its design as a final product. Only in this way can a rational approach be determined for undertaking a site plan that recognizes and responds to the seismic hazards in an area.

Prior to the start of any preliminary design drawings, it is prudent for the design professional to take the time to complete a land use assessment of the site using seismic microzonation data. In extreme cases where immediate preliminary, conceptual design approaches are required for precontract discussions with the client, it still behooves the architect to assess the earthquake hazards of the site beforehand. To do otherwise would be negligent. As the expression goes, "It is better to be safe than sorry."

CHAPTER

5 BUILDING DESIGN

Because of the many variables involved when dealing with earthquakes, it is often said that the seismic design of a building as currently addressed represents a state-of-the-art approach in earthquake engineering rather than being an exact science. In the application of lateral loads to a building system, much is done by convention, and much use is made of modifying factors and coefficients to simulate the impact of an earthquake on a building as closely and analytically as possible. It is critical, therefore, that the modeling of seismic forces as they move through the building's components follows a logical force path for the structure to be able to resist them directly without major complexities. Whenever possible, it is important to avoid eccentricities known to produce stress concentrations at undesirable points that are difficult to analyze and calculate.

STATIC AND DYNAMIC LATERAL FORCE PROCEDURES

Generally speaking, two methods are used in the UBC to determine the equivalent lateral loads to be applied to a building's structural system. The first system, used for small buildings limited to 240 ft in height and located in zones of lesser seismicity (Zones 1 and 2), is referred to as the static lateral force procedure. The second, called the dynamic lateral force procedure, is used in the design of buildings over 240 ft high located in any seismic zone, and those over five stories or 65 ft located in the higher zones of seismicity (Zones 3 and 4).

For the purposes of this text, which is oriented toward the professional architect, the principal focus will be on details of the static procedure used in the design of small buildings. The dynamic procedure will also be described in concept so that the architect will be familiar with the general terms employed and the directions required for the dynamic spectra design of high-rise buildings.

EARTHQUAKES AND FUNDAMENTALS OF GROUND MOTION INPUTS

Before undertaking the seismic design of a building system, two fundamentals must be thoroughly understood. First, it is important to realize that the so-called damaging seismic forces experienced by a building are technically not produced directly by the earthquake. This statement may come as a surprise to many, but technically, earthquakes in themselves do not *produce* the dynamic forces for which we design a building, but rather the forces are *induced* into the building by the earthquake's ground motions. An interpretation of the earthquake's ground motions is first needed to understand how a building system is expected to behave in reaction to ground shaking produced by the event.

The earthquake produces ground waves and motions which are responsible for the accelerations introduced into the building's superstructure through its foundations and which are subsequently experienced by the structural components. As acceleration alone cannot produce a force, a second factor is needed to do so. As previously indicated in Chapter 2, to obtain the magnitude of the total seismic force (base shear) acting on a building system, the following equation used in the static lateral design method is based on fundamental physics following Newton's law of gravity:

$$F = MA.$$

Accordingly, it is the combination of two factors, (1) the ground acceleration a multiplied by (2) the mass (total weight, or total seismic dead load) M of the building system (as a product of gravity, $1g$), that produces the force that places the building's components under stress. In other words, if we do not have a mass to begin with (i.e., if $M = 0$), there wouldn't be any resultant forces to act on the building even if a magnitude 8.6 earthquake struck. We need the mass (weight of the building as a product of gravity) for the acceleration to act on in order to produce a dynamic force. As a result, one of the first seismic design principles to be learned is that the lighter the building mass, the less the resulting design force will be. It is for this reason that the rule "Avoid unnecessarily massive, heavy roofs and construction when possible in earthquake hazardous areas" is often cited as a fundamental principle in the seismic design of a building.

The second fundamental in earthquake engineering is that no matter how complex a design process is, the final, governing standard in seismic design rests on the successful performance of the entire building system during the earthquake, as opposed to the failure of a part. The entire building system must hold together as a whole unit. However, as in many such cases seen in the field, partial failures, even those limited to the rupture of one critical element of the building system, are not tolerable in the ultimate sense if they lead to a progressive failure pattern and/or total collapse.

PHILOSOPHY OF BUILDING CODES

The bottom line of a successful design comes down to the fact that failures due to the overstressing of any *one critical* element in the building should not occur. The ultimate strength of any critical material used in the basic construction of a building system must be respected at all times whether it be wood, plastic, metal, concrete, or masonry. It is the capacity, or strength, of a material to resist stresses without failure that becomes the ultimate standard. Because of this, it is often said that "during an earthquake, a building is only as strong as its weakest link." It is also for this reason that many structural engineers speak about the necessity of "redundancy" when addressing the complexities of the state-of-the-art aspects in the seismic design of a major large-scale building.

Finally, in earthquake-resistant design, it is clearly understood that minimum performance standards are manifest in building codes. What is less understood is the fact that seismic provisions in most building codes currently in use are intended to protect life and reduce property damage. The key to the above sentence is the word *reduce* rather than *eliminate.* In effect, what is being recognized by establishing minimum standards in current building codes is that it may be generally unattainable in economic terms, if not impossible in practical terms, to attempt to build an "earthquake-proof" building capable of resisting all major seismic events regardless of magnitude (refer to Chapter 2). At the other end of the scale, however, it is anticipated that structures designed under lateral load requirements in building codes following recommendations by the Structural Engineers Association of California (SEAOC) should be able to (1) "resist minor earthquakes without damage" and (2) "resist moderate earthquakes without structural damage but with some nonstructural damage."

Accordingly, while building failure leading to life loss may not be acceptable, some building damage is to be expected in a major earthquake even for those buildings designed under current building code standards. Minimum code standards do not necessarily include damage control under all circumstances. It is most prudent for the architect to explain this code philosophy thoroughly and clearly to the client quite early in the process to avoid any misunderstanding later. This basic philosophy must be understood by all parties involved in the building process. At the end of such a discussion, the client may wish to place more emphasis on damage control over and above the minimum standards required by the building code. This option may be proposed for buildings housing high-tech products or for emergency service facilities.

BASIS FOR THE SEISMIC DESIGN OF A BUILDING

The seismic design of a building, no matter how elementary or complex, simply means that we have to take into consideration all of the earthquake and building factors presented up to this point in the text and pull them together in a logical synthesis. In addition to supporting all the normal static vertical loads produced by gravity, our building must now be given the capacity to resist dynamic lateral loads (seismic loads) induced into the building foundations through ground shaking. This shall be done in this book by using the current seismic provisions, Chapter 23, in the 1988 edition of the Uniform Building Code (UBC). It will be the starting point, as in its application the code assumes a very logical approach in using as many factors as possible to quantify the critical performance characteristics of a building responding to earthquake motions.

Although earthquake engineering research is typically ahead of insitutionalizing building code provisions, changes to building code standards are the result of practical applications of research findings. As current research efforts result in a better understanding and modeling of earthquakes in the fields of geology and seismology, findings are incorporated into the building code. For example, changes made in the seismic provisions of the 1988 UBC were influenced by a better understanding of the (1) role of building configuration is seismic performance, (2) earthquake risk on a regional basis, (3) effects of soils on earthquake ground motions, and (4) performance levels of the basic lateral-load-carrying structural system.

UNIFORM BUILDING CODE DEFINITION OF BASE SHEAR

The basic equation used to determine the total seismic load (base shear) to be carried by the building remains, as indicated earlier, is

$$F = MA.$$

Unfortunately, however, the elements contained in this equation as written are not sufficient to come to a rational calculation of the magnitude of the base shear F as a result of a specific earthquake event. For practical application the equation must be rewritten to fit conditions applicable to building systems.

While it is a simple process to calculate the mass M (total weight or total seismic dead load) of a given building once we know its configuration, layout, and materials of construction, there is no clue as to what value should be used for the acceleration a simply because history indicates that one earthquake event may have different characteristics than another, depending on magnitude for one thing, even when triggered by the same fault. Therefore, the global value of the experienced acceleration a will also vary depending on the magnitude of the earthquake and the geophysical characteristics of its location. Plainly, as written above, the equation does not contain enough elements for us to take into account other variables that allow us to approximate a better modeling of a building's reaction to an earthquake.

What other variables should be included in the equation? For starters, what role should the seismicity of a given region play in the equation? Everything else being equal, should a building in a zone of very low seismicity be designed under the same standards as one located in a zone of extremely high seismic risk? Of course not! Otherwise, for economic reasons, if nothing else, one would quickly come to the conclusion that the building located in the zone of low seismicity would be far overdesigned. Consequently, we should add a modifier to our equation in order to reflect the level of risk assumed when constructing a given building in a specific seismic zone. To make it simple, we will give that modifier the symbol Z for its actual location in a zone descriptive of the site's seismic risk. So now we can rewrite the equation with the added element included as follows:

$$F = ZMa.$$

What other variables should we consider in modeling the response of a building to an earthquake? We know that there are several structural systems that may be selected for the seismic design of a building using different materials of construction; for example, reinforced concrete frame, steel braced frame, shear wall, and mixed construction, among others. It is also known that each system will have different performance characteristics in response to an earthquake. In order to recognize this distinction, therefore, we should add an element to the equation that recognizes a building's reaction to an earthquake event depending on the type of structural system used to resist lateral loads. Let's identify this factor by adding the symbol K to our equation:

$$F = ZKMa.$$

Another modifier that should be added to the equation is related to the significance that we assign to a building's function as an expression of its importance to society in surviving the earthquake.

For example, should the survival of a nuclear power plant during the earthquake be considered of equal importance to societal needs as a detached wood-frame carport? Of course not.

Over the years, therefore, we have clearly accepted the fact that one building type should be designed to a higher standard from another depending on its importance and/or occupancy. On a national level this occurs in the design of nuclear facilities, failure of which could be catastrophic to humanity. At a state level, in California, after the 1933 Long Beach and 1971 San Fernando earthquakes, legislative action resulting in passage of the Field Act in 1933 and the Hospital Act in 1972 required that both public schools and hospitals be designed to higher seismic performance standards than other building types. In recognition of this principle, the seismic code requires that the significance of a building in terms of societal needs be specified in the equation. This modifier is entered into the equation as the symbol I for building importance. At this point the equation now looks like this:

$$F = ZIKMa.$$

We are now ready to tackle the problem that revolves around the task of finding the best way to define the ground acceleration expressed by the factor a. This proves to be very difficult, since, as indicated earlier, there are an infinite number of accelerations associated with the numerous earthquakes experienced over the years. How to model this in a simple manner that is comprehensible and technically acceptable becomes an almost impossible undertaking. It is at this point that technical experience and professional judgment play an important role in realistically interpreting and expressing the factors that best define the variables associated with ground accelerations transmitted to the building. Here, more than any other place in the promulgation of code standards, state of the art enters into the picture to determine the equivalent static lateral load to be applied to a building system.

One more step is needed at this point to determine the factors that best define those building variables that optimally describe a building's basic structural system, its ductility, fundamental period, and site–structure interaction in response to ground-motion accelerations. For practical application, these four factors, more than any others, have been deemed to be the most descriptive of a building's response to the ground accelerations a caused by the earthquake. As seen, their symbols are respectively expressed as K, C, T, and S.

In the 1986 UBC, these factors were taken into consideration and the base shear equation was expressed as

$$V = ZIKCSW.$$

In this equation, K is the factor based on the structural system used to resist earthquake lateral loads, while the the symbol C is a combi-

Figure 5-1 Damaged medium-rise building (note damaged elevators at bottom of service core), 1985 Mexico City earthquake.

nation of the building's period and a soil–structure interaction response factor to ground accelerations.

However, after the 1985 Mexico earthquake, which severely damaged medium-rise buildings in Mexico City (see Figure 5-1), and additional data analysis completed on building performance in the United States by the SEAOC, the equation used to determine the total base shear was further changed in the 1988 edition of the UBC relative to the performance of basic structural systems used to resist earthquakes. (Note: Experiences from the 1985 Mexico earthquake also led to the addition of a fourth soil type, S4, in 1988 UBC Table No. 23-J, "Site Coefficients.") By convention and for practical application purposes, the equation used to determine this total equivalent static lateral force V (base shear) as formulated for the building code reflecting these changes is now rewritten in the following form:

$$V = ZIC/R_w \times W.$$

Example Problem: Design of a Simple Building

In this example, a small wood-frame building located on the West Coast will be designed to resist earthquake forces following the static lateral load procedure outlined in the 1988 UBC seismic provisions. In this building design problem dealing with a light wood-frame structure, plywood shear walls will be used as the lateral-forces-resisting structural element.

Program requirements are to design a one-story wood-frame commercial building, 1800 square feet in size, to serve as a computer software retail outlet. The store will be located in a part of the Silicon Valley in the San Francisco Bay Area.

As a result of the preliminary design process, the building's layout and configuration, in answer to the programmatic requirements, have been determined as indicated in Figure 5-2. The client has accepted the preliminary design, so the next step is to design the structural system with a capacity to resist the earthquake forces anticipated in the area, according to the 1988 UBC's seismic requirements. The procedure now involves determination of the components of the base shear equation ($V = ZIC/R_w \times W$) in order to obtain the value of V, the total base shear acting on the building, by taking the following steps:

1. The location of the site is in the San Francisco Bay Area, which is in Seismic Zone 4 according to the U.S. seismic zone map shown in Figure No. 2, "Seismic Zone Map of the United States," of the UBC. Referring to Table No. 23-I, "Seismic Zone Factor Z," the value of Z in Zone 4 is 0.40.
2. Referring to Table No. 23-L, "Occupancy Requirements," in the UBC, the importance factor I for a standard occupancy structure IV such as retail store is given a value of **1.0.**
3. The value of C, a numerical coefficient related to the horizontal force factor, is found as follows:

   $C = 1.25S/T$ where S is the site coefficient for soil characteristics given in Table No. 23-J, and T is the fundamental period of vibration, in seconds, of the structure in the direction under consideration.

   However, the UBC also specifies that the value of C need not exceed 2.75 and may be used for any structure without regard to soil type or the structure's period. (An exception to this provision is made where code prescribed forces are scaled up by 3 [R/8], in which case the minimum value of the ratio C/R shall be 0.075.)

4. The final step is to determine the value of R_w which we obtain from Table No. 23-O, "Structural Systems," which in this

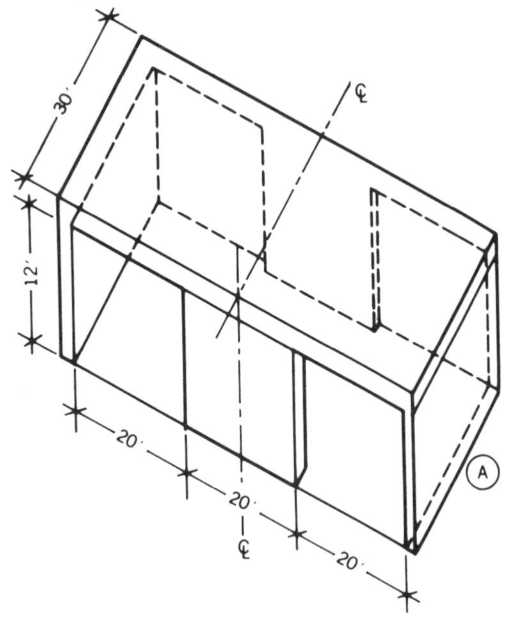

Figure 5-2 Isometric of building layout and configuration for example problem.

case is **8** for a bearing wall system consisting of light-framed walls with shear panels and plywood walls for structures of three stories or less.

5. Substituting the numerical values from steps 1 through 4 above into the total base shear equation gives us the following results when using the value of C which need not exceed 2.75:

$$V = 0.40 \times 1.0 \times 2.75/8 \times W$$

or

$$V = 0.1375 \times W$$

6. Now all that is left to find the value of V is to calculate the value of W, or the building's total seismic dead load, and multiply it by the coefficient 0.1375 indicated above.

CALCULATION OF W, THE BUILDING'S TOTAL SEISMIC DEAD LOAD

In order to calculate the building's total seismic dead load for use in the static lateral load procedure analysis process, it is first necessary to compute the weights of all parts of the total building system that contribute to the development of the lateral force to be applied to the building. All of the gravity dead loads of the building should be considered in calculating *W* for eventual base shear computations. Floor live loads are seldom, if ever, included, except for major permanent and fixed equipment and furnishings. Snow loads require careful attention and are another exception in many parts of the country that require an allowance for their inclusion. In all building codes it is assumed that lateral earthquake loads may come from any direction, so the process involves applying the resultant forces acting on the building in both axial directions, north/south and east/west (see Figure 5-3).

Before proceeding with any calculations, it is first necessary to understand another fundamental concept in earthquake-resistant design: how the component weights of a building system contribute to the development of lateral forces acting on a structure. Referring to Figure 5-3, the walls of a building system, when pushed laterally by a load composed by its inertial dead weight and the earthquake's horizontal ground motion components, are conceived by convention as spanning the vertical space between the floor/foundation line and the roof elements above. It can be assumed then that only the top half of the wall will push against the superstructure of the roof system. In similar fashion, the lower half of

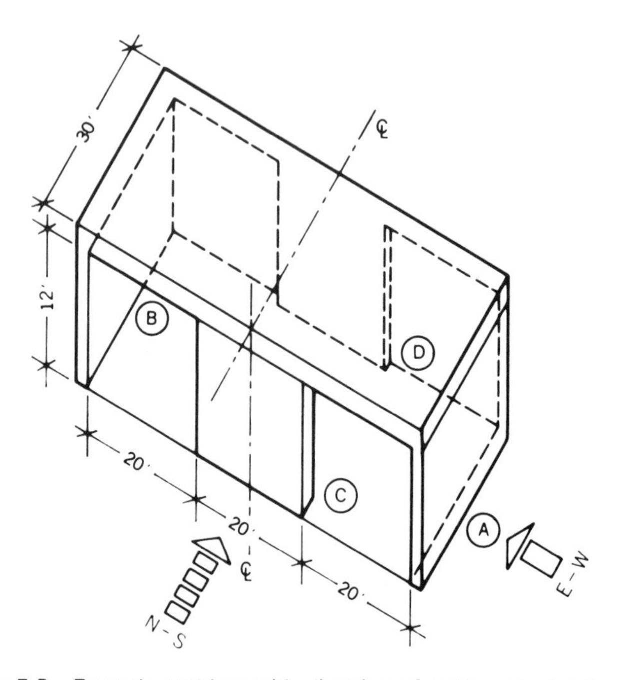

Figure 5-3 Example problem with direction of earthquake loadings indicated.

the wall will react against the floor/foundation line, and the ensuing force will be automatically transmitted into the ground directly from there. On this basis, in a one-story building, it is considered that the lateral force generated by the lower half of the wall is a passive one, which may be generally ignored as it is dissipated directly into the ground, while the load generated by the upper half of the wall is an active one, which contributes its weight proportionately to the building's total seismic dead load. (Note: In a multistory building, all the weights of the building system above the second floor, including walls, contribute their part to the total seismic dead load. It is only at the ground floor, or first floor, that the weight of the lower half of the wall is assumed to be dissipated into the ground through the foundation system and thus ignored in calculating the total seismic dead load for building design purposes [see Figure 2-3*c*].)

With the basis for this fundamental concept understood, we may now proceed with the calculation of the total seismic dead load, in which, step by step, we must take into consideration the total of all the weights of all the construction elements and components that make up the entire shell of the building system; that is, the total of all the dead load parts of the entire building shell, including walls, floors, roof, parapets, beams, girders, columns, wall piers, and so on.

To do this, we must first select the various and appropriate materials of construction that are to be used in the design of the building. As it was initially determined that the building would be designed as a light wood-frame system, construction materials representative of this building type would be selected for their appropriateness. In this case, as a representative example, the following materials are selected:

Component	Materials	Weight*
Walls	2 × 8 wood studs at 16 in. o.c.	3
	3/4-in. plywood sheathing	3
	2-in. rigid insulation	3
	7/8-in. cement stucco exterior	10
	Glass and metal frame wall exterior	12
	1/2-in. gypsum board interior	3
	Subtotal	34
Ceilings	1/2-in. gypsum board	3
Roof	2 × 10 in. roof joist at 16 in. o.c.	5
	3/4-in. plywood sheathing	3
	4 × 16 in. Glu-Lam beams at 10 ft o.c.	5
Roofing	Composition, 5-ply tar and gravel	9
	Subtotal	25

*Note: Pounds per square foot.

At this point we may calculate the total seismic dead load weight of the building:

Roof:	25#/sq ft × 30 ft × 60 ft	= 45,000 lb
Walls:	Stucco - 22#/sq ft × 6 ft* × 100 ft	= 22,000 lb
	Glass - 12#/sq ft × 6 ft* × 80 ft	= 5,760 lb
	Total Building Seismic Dead Load:	72,760 lb

*Note: Top half of 12-ft-high walls.

According to this calculation, any earthquake-resistant structural system that is used in the design of the building must develop a capacity to support the total seismic dead load of 72,760 lb. Furthermore, since the earthquake's lateral ground motions may come from any compass point, the structural system must be designed to resist the seismic load in each direction of the building's axis, north/south and east/west.

The final step is now to select an appropriate earthquake-resistant structural system, analyze it, and, based on the findings, complete the building's design.

DESIGN AND ANALYSIS OF THE BUILDING'S EARTHQUAKE-RESISTANT STRUCTURAL SYSTEM

First, the primary construction type of the building is a light wood-frame system. Second, it is a relatively simple building in size and configuration: light weight, having a symmetrical plan, one story high, and covering a small area 30 ft × 60 ft in size.

As indicated in Figure 2-3, there are three basic conventional structural systems that may be appropriately used in the earthquake-resistant design of this simple building: (1) shear wall, (2) braced frame, and (3) moment frame. As it is difficult to achieve a moment-frame-type system without the introduction of other materials in a wood-frame building, the moment-frame option is eliminated in favor of the other two. Also, the use of a base isolation system in such a simple building, with its noncritical occupancy rating, in all probability would not be warranted in terms of appropriateness or cost-effectiveness.

Again, in consideration of the first two systems, shear wall or braced frame, it is important to recognize that we are dealing with a simple wood-frame building. Fortunately, its preliminary design has resulted in four solid walls without any openings, and all are well configured for use as effective shear walls: two 30 ft × 12 ft walls in the north/south direction, and two 20 ft × 12 ft walls in the east/west direction. For the first trial analysis, the consensus would obviously be to design a box/plywood shear wall concept for its compatibility with the building's wood-frame construction type as the building's earthquake-resistant structural system.

It is crucial to select an earthquake-resistance design system that is clearly compatible with the basic construction type of the building. For example, if the construction type under consideration had been a steel frame or a reinforced concrete frame, a moment-frame system or a braced-frame system would have been equally appropriate for use as the earthquake-resistant structural design concept. In this case, because of its extreme ductility, use of a plywood shear wall system would not be suitable in a concrete or steel frame system. In fact it would not be permitted under any building code standards. However, it still would be possible to consider a shear wall system in a steel or concrete frame as long as it is constructed as a reinforced concrete shear wall system and not a plywood type.

Continuing with the design of the box/plywood shear wall earthquake-resistant concept for our wood-frame construction type, it is now necessary to decide which walls of the building are suitable for development as shear walls. Returning to the overall architectural design of the building (see Figure 5-2), it is evident that the building has four exterior solid walls of considerable length (two on each side of the building), one longitudinal set 20 ft long, and the other set at the ends, each 30 ft long. A logical conclusion, then, is to develop these four walls as wood-frame/plywood shear walls as the earthquake-resistant design system.

Having made this selection for the design concept, trial designs for the earthquake-resistance capacities of the shear walls must be undertaken. The shear walls in the north/south direction will be considered first (Walls A and B), followed by the two walls in the east/west direction (Walls C and D).

Starting with Wall A, the initial task is to determine the magnitude, or proportion, of the lateral load to be carried by the wall when it is developed as a plywood shear wall. Referring to Figure 5-4, we must first proportion the building into tributary areas related to each shear wall to determine its resistance capacity to lateral loads. Since two walls exist in the north/south direction, Walls A and B, it is only logical to divide the building into two equal tributary load areas, as indicated; thus, Wall A carries half the load and Wall B the other half, thus balancing the resistance of lateral forces between the two walls and eliminating any potential negative torsional effects due to eccentrically placed earthquake-resistant structural elements. In this design concept, in effect, each wall will be required to carry its share, or 50 percent in this case, of the base shear lateral force derived from the total seismic dead load of 75,760 lb, as calculated previously.

Following the steps previously presented, and referring to the equation by which the total base shear is determined, the following

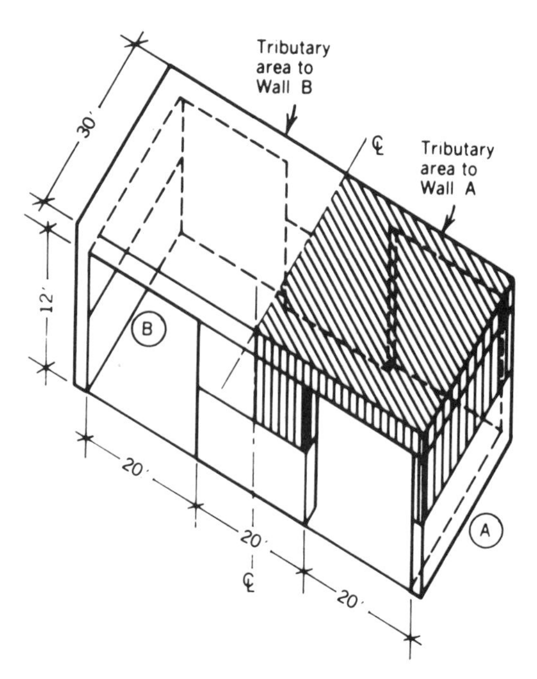

Figure 5-4 Building divided into two tributary areas.

sequence is followed in substituting numbers in their appropriate places:

$$V = ZIC/R_w \times W$$

$$V = 0.1375 \times W$$

$$V = 0.1375 \times 76{,}760 = 10{,}417 \text{ lb.}$$

As there are two shear walls in this direction, each plywood shear wall is to carry 50 percent of the lateral base shear load. Shear Wall A and Shear Wall B must be developed with a capacity to resist

$$V \text{ for Shear Wall A, or B} = 1/2 \times 10{,}417 = 5209 \text{ lb.}$$

(Note that as shown in Figure 5-5, if there had been three solid walls in the north/south direction, with the third one in the middle of the building in response to architectural program requirements, the building would then have been divided into three tributary load areas, one assigned to Wall A, another to Wall B, and the third to the center Wall E. In this case, each shear wall would be carrying a tributary load area of a lesser magnitude than was the case when only two shear walls were used at the two ends of the building. In short, each of the three walls would be required to carry less of a

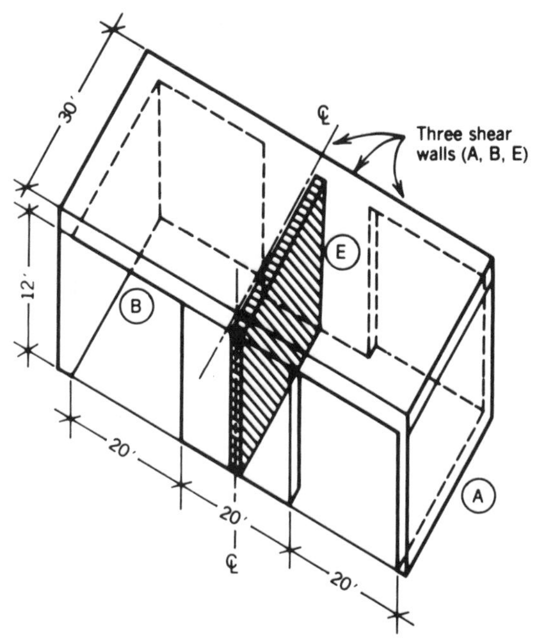

Figure 5-5 Building with three potential shear walls indicated, Wall E in center.

lateral load proportionately than the two shear walls in the building designed above.

Now that the lateral load to be carried by each shear wall has been calculated, to complete the process the construction specifications of the walls must be determined. We start by first recomputing the total lateral load to be carried by the shear wall (5209 lb) into a minimum unit lateral load to be carried per linear foot of length of the shear wall:

$$V \text{ for Shear Wall } A, \text{ or } B = 5209 \text{ lb/30 ft}$$
$$= 174 \text{ lb/ft.}$$

From this result it is determined that each linear foot of shear wall is required to carry a minimum of 174 lb. In the 1988 UBC, Table 25-K gives the maximum allowable shear wall stresses that may be developed for plywood panels. It is now a simple matter to refer to Table 25-K and select the size (thickness) and quality of plywood panel which with an appropriate nailing schedule does the job.

According to Table 25-K, a 1/2-in. thick exterior-grade "A-C" plywood panel with 6d nailing at 6 in. on center (o.c.) at the edges has the capacity to resist a load of 125 lb/ft, in comparison to others in the table which do not. Hence, for the construction of the plywood shear walls, we would use the 1/2-in. plywood with the grade and required nailing pattern specified.

It is the function of the roof system to act as a horizontal shear diaphragm to transmit the lateral loads from the center of the building into the respective shear walls. Failure to develop the roof system as an adequate horizontal diaphragm for this function simply implies that the lateral loads from the center of the building would never reach the shear wall locations and not allow the earthquake-resistant design of the structure to act as a unit. The design of a horizontal diaphragm follows the same process as the design of the vertical shear wall presented in this chapter. The steps required to calculate the loads and resistant capacity of a plywood roof diaphragm would be exactly the same.

Another detail of construction that should be specified is how the plywood joints will be handled. As a structural shear-wall-resisting element of the building, it is important that the horizontal plywood joints be staggered to ensure the integrity of the wall as a shear wall. As the wall is 12 ft high and 30 ft long, a plywood pattern with staggered 8 and 4 ft joints should be selected with solid blocking behind all joints to assure proper nailing at all edges of the plywood. Further, to avoid potential buckling of the interior field of the plywood, which could lead to a loss of shear resistance capacity, 6d nailing at 12 in. o.c. should also be provided at the interior stud lines and blocking area.

The final task is to check requirements for the connection of the plywood shear wall to the foundation. Unless this is properly accounted for, the laterally induced earthquake forces could literally slide the building off the foundation and dump it onto the ground. As shown in Figure 7-2, this has happened during past earthquakes to many older wood-frame structures that were not bolted down to the foundation.

Another significant element in shear wall design in wood-frame construction is making sure that the plywood is securely nailed to the top plate of the wall and bottom mudsill in the foundation. Only by doing so will the loads from the roof be adequately transmitted to the shear wall and those from the wall to the foundation system.

As computed earlier, the maximum design shear force on Wall A is 5209 lb. A connection system holding the wall to the foundation against lateral shear loads must be designed to carry this force; a typical system is to use steel foundation bolts for this situation. The allowable stresses for anchor bolts in single shear are listed in the 1988 UBC and the *AISC Steel Construction Handbook*. By selecting a 5/8-in. anchor bolt, a size commonly used for foundation anchors and having a capacity of 3100 lb in single shear, a minimum of two bolts would be required:

$$5209 \text{ lb total load}/3100 \text{ lb per bolt} = 1.68 \text{ bolts.}$$

However, checking with the 1988 UBC requirements, it is recommended that foundation bolts be placed at 6 ft o.c. as a minimum requirement, which over a 30-ft wall length would require a minimum of six bolts, with one at each end. As in this case, building code requirements govern the design, and six 5/8-in. anchor bolts at 6 in. o.c. would be placed in the concrete foundation as a means of connecting the shear wall and adequately transmitting the shear stresses from the shear wall into the foundation.

One last task is left: to check Shear Wall A for potential overturning. Figure 5-6 indicates several possible configurations of shear wall elements. In any given building, where we may have openings in walls for passage, doors, or windows, it is not always possible to have solid uninterrupted wall surfaces 30 or 20 ft long as in the building example in Figure 5-2, where its length is greater than its height. Some of the walls in Figures 5-6 are taller than they are long, and therefore may present problems of failure by overturning, because a simple 5/8-in. foundation bolt may be pulled out of the concrete foundation or wood mudsill owing to excessive forces. Figure 5-7 shows two examples of metal hold-downs with sufficient capacity to withstand overturning caused by a potentially excessive overturning moment owing to a less-than-desirable configuration of a laterally loaded shear wall.

This task is the last in the standard, conventional, and basic design requirements of a plywood shear wall as an earthquake-resistant structural element. To complete the overall design of the entire building, it is necessary to go through the same process for the design of Walls C and D as shear-wall-resisting elements in the east/west direction. The roof system, a horizontal shear diaphragm (to distribute the lateral loads from the center of the building to the respective shear walls), must also be checked to complete the seismic design of the total building system.

If our example building had been larger in area and number of

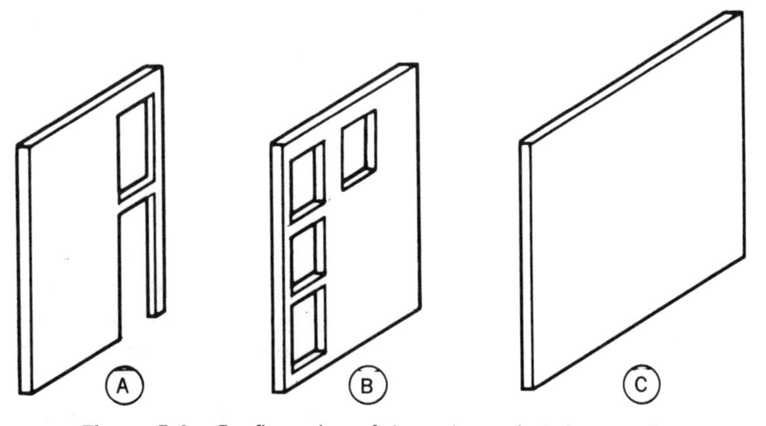

Figure 5-6 Configuration of three theoretical shear walls.

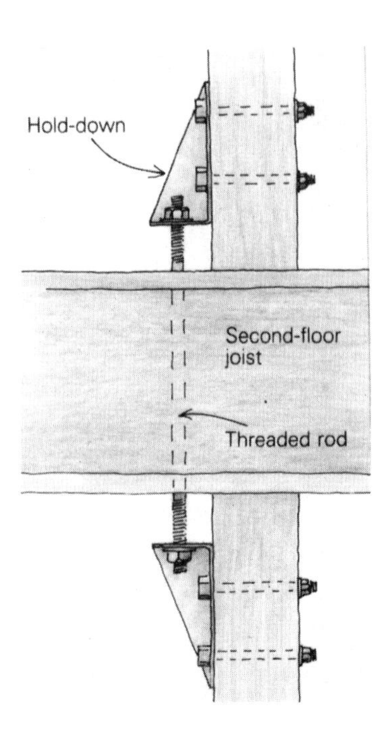

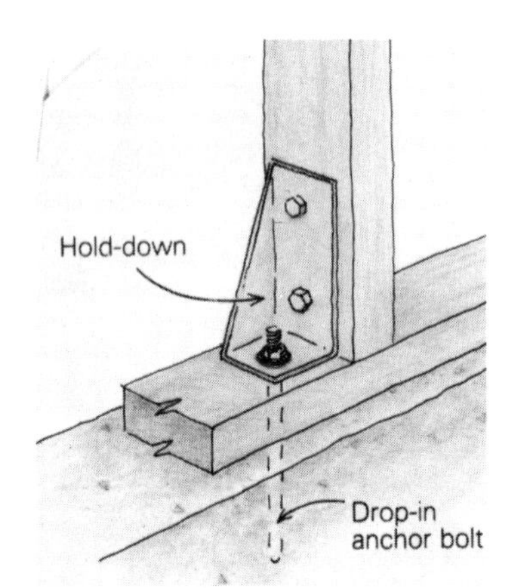

Figure 5-7 Two examples of hold-down details. *Source:* Helfant (1989). Reprinted with permission.

stories, another solution could have been to use moment-resisting steel frames to resist lateral loads instead of the plywood shear wall system. Figure 5-8 indicates the use of moment-resisting steel-frame members at the corners of a two-story wood-frame building. In both cases, however, it is critical to frame the entire structural system with appropriate collectors to carry the loads to the earthquake-resistant elements, moment frames, or plywood shear walls.

Figure 5-8 Wood-frame building developed with lateral-load-resisting steel moment frames at corners.

DESIGN OF OTHER BUILDING CONSTRUCTION TYPES

The process involved in the static lateral load procedure for the seismic design of other building construction types, reinforced concrete, steel, masonry, and mixed construction, is similar to the example for the wood-frame building above. The basic tasks are to

1. Take into account, from the very beginning of the preliminary architectural design, the implications of seismic forces by working with a structural engineer,
2. Determine the seismic lateral load coefficients for the site conditions and structural system *(Z, I, C,* and *R),*
3. Compute the total seismic dead load *W* of the building system based on the weights of its construction materials and construction type.
4. Compute the total base shear *V,*
5. Determine the basic concept for the structural seismic lateral-load-resisting system and its configuration,

6. Determine the tributary load areas related to the location of the lateral-load-resisting structural elements,
7. Design the respective lateral-load-resisting structural elements and ensure that all their units act as a whole, and
8. Review and assess all assumptions regarding and concepts selected for compliance with seismic safety objectives and public health and safety goals.

Figure 5-9 illustrates the use of steel "K" braces as seismic braces in a multistory steel-frame building in northern Calfornia.

The bibliography at the end of this book contains references that the reader may find useful for the seismic design of all types of buildings. Because of the nature and purpose of this text, and because the process is similar in the majority of cases, it is not the intent of this publication to present in detail the seismic design procedures for all types of buildings.

Three excellent sources for information on the design of other earthquake-resistant structural systems are the referenced publications by (1) the Building Seismic Safety Council (BSSC), *NEHRP Recommended Provisions for the Development of Seismic Regulations for New Buildings* (1988); (2) *Guide to Application of the NEHRP Recommended Provisions in Earthquake-Resistant Building Design* (Conner et al., 1987); and (3) Department of Defense Tri-Services Seismic Design Committee, *Technical Manual: Seismic Design for Buildings* (1982). These documents present detailed approaches and calculations for the earthquake-resistant design of wood, reinforced concrete, masonry, and steel-frame buildings.

Figure 5-9 Steel K braces in steel-frame building under construction.

THE DYNAMIC LATERAL FORCE LOAD PROCEDURE

Although it is not in the purview of this publication to delineate a detailed analysis illustrating the use of the dynamic lateral force load procedure, its fundamental implications are presented here so that the reader may be familiar with some of its objectives, hypotheses, and technical requirements. Suffice it to state that if an architect is involved in the seismic design of a major multistory building requiring dynamic analysis, immediate consultation with a structural engineer with detailed experience in dynamic load procedures is paramount even before any preliminary conceptual design drawings are begun.

In contrast to the static lateral force procedure presented in detail above, the dynamic lateral force load procedure for the seismic-resistant design of building systems is founded on the principles of structural dynamics. According to the 1988 UBC, the analysis "shall be based on an appropriate ground motion representation and shall be performed using accepted principles of dynamics." As a general rule, all multistory buildings 240 ft or more in height, and structures located in Seismic Zones 3 and 4 over five stories or 65 ft in height, not having the same structural system throughout (except as permitted in special situations) require a dynamic analysis rather than a simple static one.

The important structural characteristics of a building system that require dynamic analysis are determined by its

1. Natural period of vibration,
2. Ductility, and
3. Damping properties.

The fundamental considerations in the 1988 UBC that govern dynamic analysis include concerns for

1. A response spectra,
2. An elastic design response spectra developed for the specific site,
3. Ground motion time histories for the specific site representative of actual earthquake motions, and
4. The vertical component of ground motion scaled as a two-thirds component of the corresponding horizontal accelerations.

In this dynamic process, an elastic dynamic analysis of a structure is required that utilizes the peak dynamic response of all building modes (natural periods of vibration) that make a significant contribution to the total response of the structure. The required time history analysis is to contain the analysis of the dynamic response

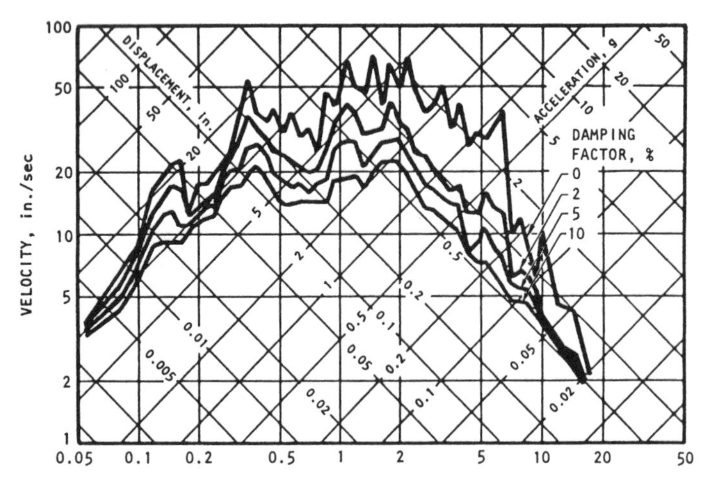

Figure 5-10 Response spectrum plot, 1940 El Centro, California, earthquake. *Source:* EERC and Anil K. Chopra.

of a structure to each increment of time when the base is subjected to a specific ground motion time history. In the process, a mathematical model of the physical structure is completed to represent the spatial distribution of the mass stiffness of the structure to the extent required for the calculation of the critical characteristics of its dynamic response. The dynamic analysis must also take into account all negative possibilities for detrimental torsional effects on the structure, if any.

The principal purpose in utilizing a response spectrum approach (see Figure 5-10) in the dynamic design of an earthquake-resistant structure is to be able to take into consideration the actual response of a building to ground accelerations by quantifying the following factors:

1. Velocity in inches per second,
2. Acceleration in feet per second,
3. Displacement in inches, and
4. Period in seconds.

In general, it is only feasible to complete a dynamic analysis of a given structure through the utilization of computers. Response spectrum curves are also replotted to adjust them to damping factors induced into the building system by various components such as architectural, nonstructural, and structural elements. Figure 5-11 illustrates normalized response spectra shapes, reflecting the influence of soil types.

Damping is defined as a factor that attempts to measure the incidence of decline in the amplitude of vibrations of a structure owing

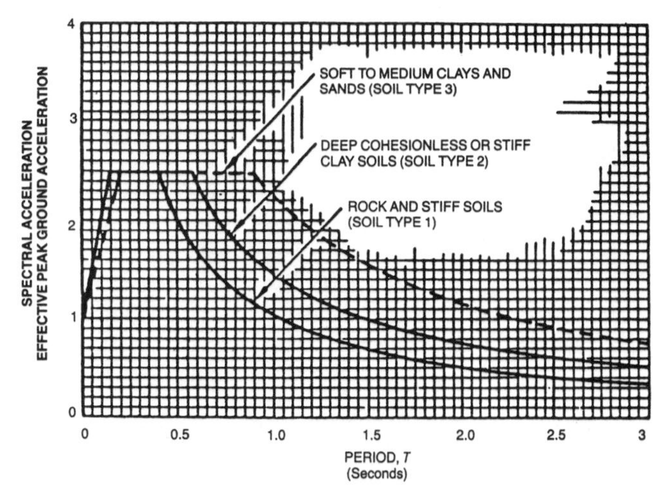

Figure 5-11 Normalized response spectra shapes. *Source:* Reprinted with permission of International Conference of Building Officials.

to the effects that internal friction and the capacity of the building system to absorb energy have on the basic structure's actual performance.

As may be seen from this brief explanation, the dynamic analysis process can become very complicated when complex multistory building systems are involved. It requires much technical knowledge on the part of the user, and above all its success depends on professional judgment and extensive field experience on the part of the structural engineer.

CHAPTER

6 NONSTRUCTURAL BUILDING ELEMENTS

The bottom line in evaluating a well-constructed building is found in its success in providing for the safety and comfort of its occupants. After assessing building damage in Anchorage after the 1964 Great Alaska earthquake, a mechanical engineer, J. Marx Ayres, made this observation relative to the occupants of buildings: "If, during an earthquake, they must exit through a shower of falling light fixtures and ceilings, maneuver through shifting and toppling furniture and equipment, stumble down dark corridors and debris-laden stairs, and then be met at the street by falling glass, veneers, or facade elements, then the building cannot be described as a safe structure" (Ayres et al., 1973). In that earthquake, at least three deaths, numerous injuries, and considerable panic were directly attributable to the failure of nonstructural elements.

A contemporary, well-designed building represents a great complexity of systems brought together through its nonstructural elements and joined into a single, operational environment; in general terms, these nonstructural elements are conceived as being separate and independent of the structural system. The basic structural arrangement of a building provides the supporting skeleton or shell of the building system, whereas the nonstructural elements provide the fabric or cladding that encloses the building exterior and defines its interior spaces.

As the needs of society and clients increase, to meet demands of a highly advanced, technically oriented culture, buildings will become even more complex. Owing to these growing complexities and demands, it is widely agreed that many more things can go wrong in a contemporary building than one designed only 10 years ago. This is particularly true when we consider all the critical nonstructural elements and systems upon which our contemporary buildings rely for viability and endurance.

EARTHQUAKES SEEK THE WEAK LINK

Through the years, earthquakes have earned a growing reputation for their consistent propensity to find the "weak link" in a complex system and lead that system into a progressive failure mode. As a result of this ability to locate and strike the weakest point of an assembly, it is often observed that earthquakes "love" complexities. The more complex a system, the more determined major earthquakes seem to be in quickly discovering, and damaging, the most vulnerable element. If that particular building element becomes a critical one owing to a concentration of forces placed on it during an earthquake, a domino effect, or progressive failure mode, may occur, and failure of many interdependent components may result. In critical cases, even total collapse of the entire building occurs.

In the seismic design of buildings, life safety and damage control are the fundamentally shared responsibilities of design professionals. Because of (1) the greater complexities found in contemporary buildings and (2) their dynamic, nonlinear performance when responding to major seismic events, these design responsibilities must be shared with all parties involved in the planning and design of earthquake-resistant building systems. When the complexities of all nonstructural elements in a building system are taken into consideration, it is obvious that the undertaking is simply too complicated to be handled by one individual.

Fortunately, a significant data base of technical information is being developed for the architect to use in the design of seismic-resistant buildings, and all design professionals should be familiar with it. Preliminary design phases of all buildings located in severe environments must have objective contributions from all members of an informed design team as early as possible. When dealing with the seismic design of many nonstructural elements in a major building, the architect, structural engineer, mechanical engineer, electrical engineer, acoustical engineer, and fire-control engineer must collaborate closely. This is especially important when designing for regions of high seismic risk.

BACKGROUND INFORMATION

As an illustration of the growing complexity of buildings from the earthquake hazards reduction point of view, a building is now identified as having three principal components critical to seismic safety concerns: (1) the basic structural system, (2) its nonstructural elements, and (3) its contents. Following a major, destructive earthquake, in many cases it may cost, depending on the severity of damage, appreciably more to repair a building's nonstructural

Figure 6-1 Damage to glazed terra-cotta screen on exterior stairway, 1985 Mexico City earthquake.

damage than its structural system damage. Examples of nonstructural damage due to earthquakes can be found in Figure 6-1.

In some publications the nonstructural building elements have also been referred to as "architectural" elements, as distinct from the "mechanical" and "electrical" elements. On other occasions they have been referred to as "secondary structural elements" in recognition of the fact that all building elements must have inherent structural properties in order to maintain their own integrity. The argument is that nothing on this earth is "nonstructural," for if it were, it would collapse by not even having the capacity to support its own weight. It even has been proposed by architect Marcy Wang (1987) that the word "extrinsic" be used instead of "nonstructural" as a means of describing building elements such as those that are not "intrinsic" to the basic structural system but that may influence its performance or in turn are impacted by it. Yet another definition by Professor V. Bertero of the University of California proposes that they be called "nonintentionally structural elements," since they are not part of the basic structural system yet assume inherent structural characteristics when a building is subjected to dynamic lateral loads.

These efforts in striving to postulate the appropriate term are most laudable and have a point. In any event, no matter what terminology is used, the objective is the same: an attempt to clearly define those elements that in the traditional sense are not part of the basic structure but that are most important to the function and operation of the total building system. Accordingly, for purposes of simplicity, the term "nonstructural" will be used in this text to describe the building elements that are the subject of this chapter, as distinct from the basic structural system of a building.

HISTORIC PERSPECTIVE

Until about the 1960s, principal efforts in earthquake hazards reduction interests and research dealing with a building's seismic design and technology were mainly the domain of seismologists, geophysicists, and civil/structural engineers. To minimize the threat to public safety, the initial goal was simply to preclude catastrophic collapse first, and severe structural damage second. Generally, during this early period, less attention and focus were placed on the performance of nonstructural elements of a building than on the earthquake-resistant design of the structural system. However, as advances in structural design and seismic analysis were achieved, improvements in a structural system's capacity to resist earthquake forces were soon realized. At this point the seismic performance of other building components assumed greater importance. Once it became apparent that nonstructural elements could be critical to the overall performance of a building in terms of life safety, damage control, and functional impairment, research on their seismic behavior started in earnest. Several architects and mechanical engineers started the investigation of this topic area in the late 1950s to identify and quantify scientifically the importance of the seismic performance of nonstructural elements as a critical factor in seismic design.

By the time of the 1971 San Fernando earthquake, it became quite clear that damage to nonstructural elements and building contents could result in serious casualties, severe building impairment, and major economic losses even when structural damage was not that significant. It was also discovered that severe damage to nonstructural elements during a major earthquake, such as the 1964 Alaska earthquake, could account for up to 65 to 70 percent (upper limit) of a building's replacement cost. At this level, even though the structural system could be relatively intact, building repair and rehabilitation efforts would be seriously questioned.

As another measure, the portion of property loss traceable to the damage of a typical building's contents was estimated on average to reach about 35 percent of the total loss. Today, even more dramatically, replacement of the damaged high-tech electronic equipment found in many buildings, such as those located in California's Silicon Valley, would cost more than the building itself. As illustrated by such specific cases, in dealing with earthquake-resistant design and damage control one must realize that a building's contents may be worth more than the building.

Because of the growing importance placed on nonstructural elements and building contents, experimental testing of representative building components, such as storage racks, lighting fixtures, and other parts, has occurred. For example, there has been, for the first time, full-scale laboratory testing of exterior cladding materials and interior partitions to determine their performance in accommodating seismic deformations of a steel moment-resistant frame under interstory drift limitations designated by the Uniform Building Code (UBC); this testing took place in 1984 at the Building Research Institute in Tsukuba, Japan, as part of the United States–Japan Natural Resources Program. The results of the laboratory tests are discussed in detail later in this chapter.

CLASSIFICATIONS OF NONSTRUCTURAL ELEMENTS

Nonstructural elements of a building include all those parts of the total building system and its contents that are not part of the basic structural system, which is defined by: (1) vertical support components (columns, piers, bearing walls, foundations, etc.); (2) horizontal span members (floor slabs, beams, girders, rafters, truss, space frames, etc.); and (3) any other structural element used for supporting the building's basic live loads and dead loads. The basic structural system of a building is designed to withstand all static live and dead loads, as well as all dynamic loads such as winds and earthquakes, without any assistance from the nonstructural elements, which are predominately inserted into the building during the final stages of construction.

Nonstructural elements include all the architectural components found in a building system (e.g., cladding, ceilings, partitions, doors/windows, stairs, furnishings and equipment, contents, parapets, canopies, etc.), in addition to all mechanical, electrical, and plumbing components (e.g., elevators, lights, piping, ducts, HVAC systems, escalators, security systems, fire protection systems, telephone and communications systems, computer equipment, etc.), whether on the exterior or interior. The perspective diagram in Figure 6-2, from Robert Reitherman's publication, ***Reducing the Risks of Nonstructural Earthquake Damage: A Practical Guide***

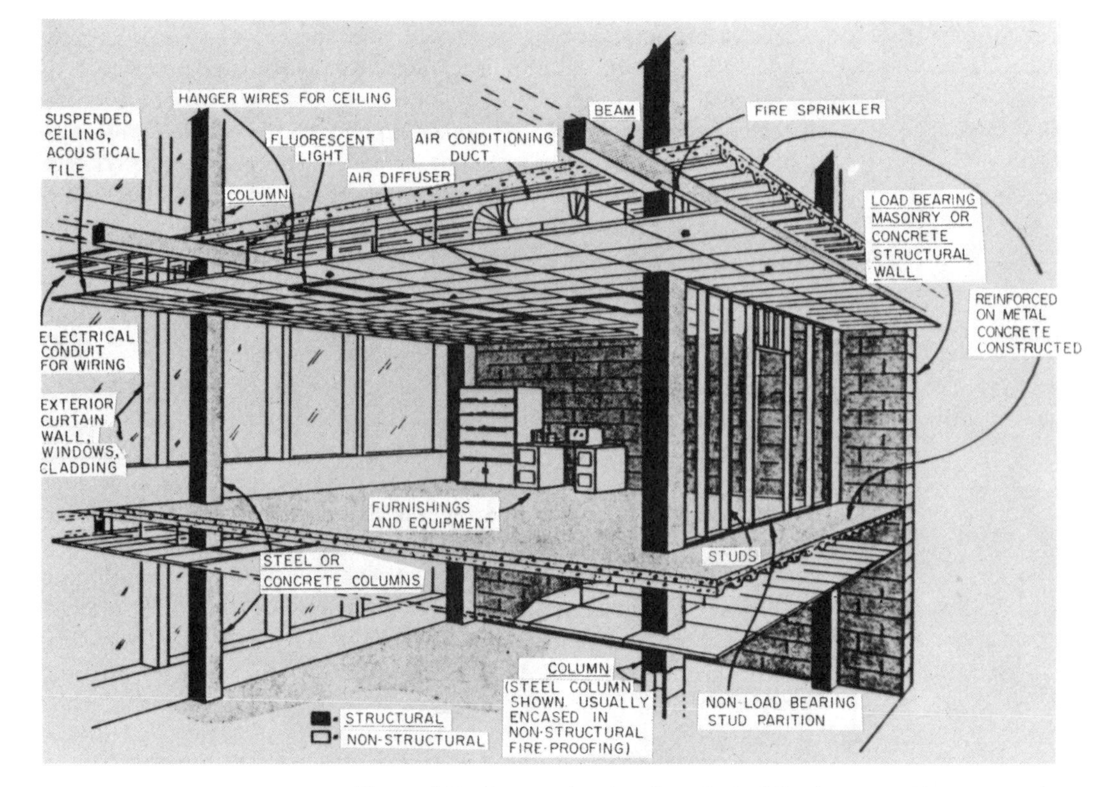

Figure 6-2 Perspective drawing of a building's structural system and nonstructural components. *Source:* Robert K. Reitherman.

(1983), illustrates the nonstructural elements of a typical building as distinct from the basic structure. A more thorough sampling of the principal categories of nonstructural elements is indicated in Table 6-1.

DAMAGE TO NONSTRUCTURAL ELEMENTS

Damage to nonstructural elements of a building is generally caused in two ways:

1. By differential movement and distortion of the primary structure, and
2. By the shaking and overstressing of the elements themselves, either in-plane or out-of-plane.

Distortion-related damage may occur to any nonstructural elements when it is not able to withstand or adjust to loads caused by the deformations and deflections of the basic structure. Stiff, brittle infill walls, curtain window–walls rigidly fixed between structural components, continuous stairways between several floors, or in-

TABLE 6-1 Representative Categories of Nonstructural Elements

1. Exterior Elements:

   Cladding, veneers, glazing, infill walls, canopies, parapets, cornices, appendages, ornamentation, roofing, louvers, doors, signs, detached planters, etc.

2. Interior Elements:

   Partitions, ceilings, stairways, storage racks, shelves, doors, glass, furnishings (file cabinets, bookcases, library stacks, display cases, desks, chairs, tables, lockers, etc.), ornamentation, detached planters, artwork, etc.

3. Mechanical/Electrical/Plumbing Elements:

   HVAC equipment, elevators, piping, ducts, electric panel boards, life-support systems, fire protection systems, telephone/communications Systems, motors/power control systems, emergency generators, tanks, pumps, escalators, boilers, chillers, fire extinguishers, controls, light fixtures, etc.

4. Contents:

   Electronic equipment, data-processing facilities, medical supplies, blood bank inventories, high-tech equipment, hazardous and toxic materials, antiques/fine arts (museums and art galleries), office equipment, radios, life-support equipment, etc.

Source: H. J. Lagorio *Architectural and Nonstructural Aspects of Earthquake Engineering,* University of California at Berkeley, Continuing Education in Engineering, Extension Division, July 1987.

flexible pipe risers between two or more floors are examples of nonstructural building elements at high risk to damage caused by extreme building movements during a major earthquake. Or, in another example, a full-height partition element may be crushed between floors owing to excessive interstory drift of the basic structural system.

Shaking-related damage is basically caused by the inability of an element to respond well to overall general shaking or the vibratory motions induced by the primary structure. Failure will occur in a nonstructural element when overstressed while vibrating internally, overturning, sliding, oscillating, and/or swinging back and forth.

RELATIONSHIP OF NONSTRUCTURAL ELEMENTS TO BASIC STRUCTURE

Basic engineering analyses indicate that the degree to which buildings are horizontally flexible under lateral seismic loads depends on their construction type and configuration. This flexibility is defined by the amount of interstory drift between floors; that is, when one floor drifts horizontally in relation to an other under excitation by earthquake ground motions. Generally speaking, a tall, multistory building is therefore more flexible by definition than a short squat one, but it is important to know that both will deflect laterally to one degree or another.

In addition to its own capacity to resist seismic forces without shattering or its anchorage pulling out, each nonstructural element, particularly exterior cladding and curtain walls, must also accommodate this flexibility induced by dynamic loads or be seriously damaged. Herein, of course, lies the crux of the problem in the earthquake-resistant design of nonstructural elements. This becomes of critical concern when dealing with typically slender high-rise contemporary buildings of 30, 50, or more stories. If in its structural response such a particularly flexible building approaches the inelastic range, the nonstructural elements may be destroyed in a "snap, crackle, and pop" manner. This is one of the reasons why building code planners are conscious of the need to establish drift controls even though codes may not be specifically concerned in detail about nonstructural damage except for emergency critical facilities. (Refer to Figure 6-3.)

It is further known that relatively rigid nonstructural components can and do, at times, adversely affect the dynamic characteristics and seismic response of the basic structure when they are placed into the building system without any regard to their influence on structural performance. Ill-placed heavy/rigid nonstructural elements may introduce damaging torsional effects into the

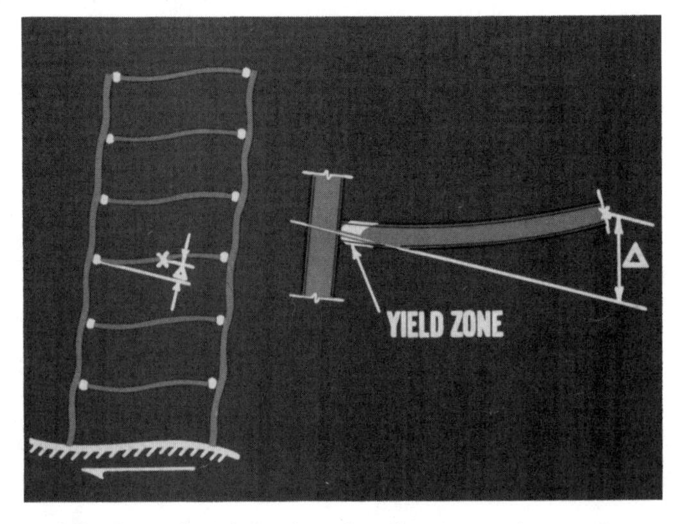

Figure 6-3 Dynamic deformation of multistory steel-frame building under earthquake lateral loads. *Source:* U.S. Geological Survey.

structural system. In this way, improper placement of nonstructural elements may negatively affect the seismic performance of a building's basic structure.

Another critical issue is related to the tolerance of nonstructural elements to motions induced in the upper stories of high-rise structures, where oscillations are greatest. An indication of extensive nonstructural damage that we have all seen is where windows have popped out of buildings not even subjected to seismic loads; the John Hancock Building in Boston is one such example. Although the induced drifts may not impose a critical threat to structural collapse, the cost of repair of the nonstructural elements, loss of building function, and potential inaccessibility to records and financial statements may have far-reaching implications. While it is generally acknowledged that drift must be controlled to acceptable levels, the proper level has not yet been clearly defined.

1984 Morgan Hill Earthquake and Structural/Nonstructural Relationships

The 1984 Morgan Hill earthquake gives some important data on the relationship of nonstructural damage to building flexibility. The performance of three buildings, in which accelerometers had been installed, provided excellent data, since each had a different structural type which made it interesting to contrast seismic responses. The three building types represented are: (1) 13-story, steel moment frame; (2) 10-story, reinforced concrete (RC) shear walls/moment frame; and (3) 10-story, RC shear walls. (See Table 6-2.) Although the structures are diverse in the number of floors and configuration, some relevant comparisons are plausible even when differences in site-specific ground motions and building heights are acknowledged.

The records from Building No. 3, steel moment frame, indicated more than 80 sec of significant structural motions in response to

TABLE 6-2 Building Data on Three Case Studies From the April 24, 1984, Morgan Hill, California 6.2 Magnitude Earthquake

				Peak Acceleration (g)	
Building	Structural System	Number of Stories	Elevation (ft)	Horizontal Ground	Building
No. 1	RC shear walls	10	96	0.60	0.22
No. 2	RC shear walls and moment frame	10	124	0.60	0.22
No. 3	Steel moment frame	13	187	0.40	0.17

Source: Mahin (1987). Reprinted with permission.

the earthquake. Substantial shifting of furniture, overturning of bookcases, movement of equipment, and drifting of other nonstructural elements occurred in this building. Peak roof displacements were on the order of 0.3 percent of the structure's height, well within UBC performance standards for elastic design. The records also indicated that there were over 20 cycles of response, with at least half at the peak response value.

The two other buildings were reinforced concrete, one with closely spaced shear walls in both directions (Building No. 1), and the other with a combination of shear walls in one direction and moment-resisting frame in the other (Building No. 2). Figure 6-4 compares the displacement path of the centroid of the roofs of all three buildings. It is generally noticed that the more rigid shear wall buildings reduce the peak structural displacements to about one-third of the value for the steel moment frame, even making

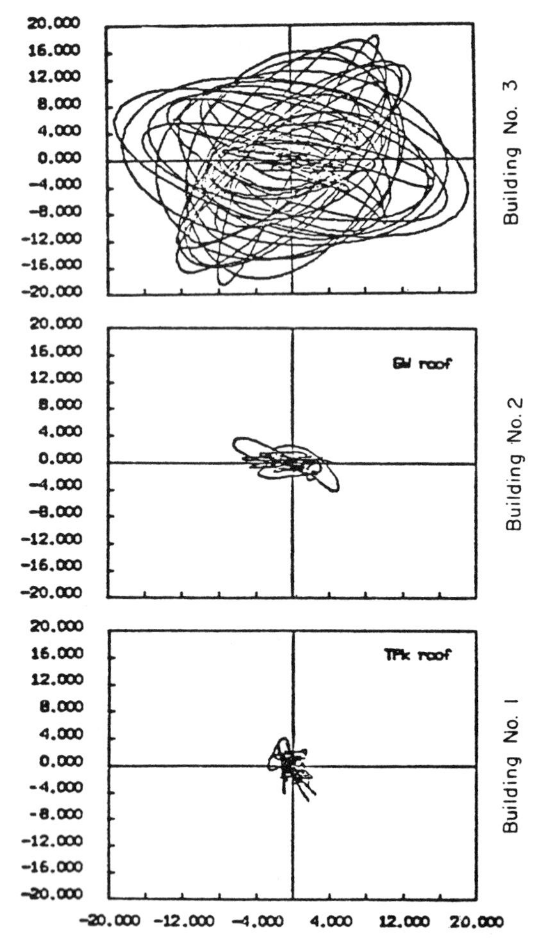

Figure 6-4 Displacement path of three buildings at roof level. *Source:* Mahin (1987). Reprinted with permission.

allowances for the increased height. The implications of these records for nonstructural damage is clear.

Exterior Cladding

One of the most vulnerable nonstructural elements in a building is the material used to clad its exterior. This exterior cladding must accommodate all building movements without failing. Torsional effects are particularly hazardous to nonductile exterior panels attached to the building frame. A classic example of the failure of exterior panels occurred at the five-story reinforced concrete J.C. Penney Building during the 1964 Alaska earthquake. The heavy precast concrete panels contributed to the torsional stresses developed by asymmetrically placed shear walls in the building and fell into the street during the earthquake. Two people were killed when the full-length, heavy panels fell onto parked cars. (See Figure 6-5.)

In some cases it is best to have a flexible curtain wall–window assembly used as exterior cladding, as its flexibility may be more compatible with the movements anticipated in flexible structural framing systems. However, one must distinguish between the performance of "stick" systems and "unit" systems and detail them appropriately. The vertical components of the stick system are typically designed to span over two floors, which implies that they must be detailed and assembled to take on considerable stresses in

Figure 6-5 Cleanup of damaged exterior cladding of precast panels on J.C. Penney Building, 1964 Alaska, Anchorage, earthquake. *Source:* K. V. Steinbrugge, private collection. Reprinted with permission.

Figure 6-6 Shattered plate glass on building front, 1987 Whittier-Narrows, California, earthquake.

accommodating interstory drift. If the system cannot accommodate this drift, failure will ensue.

The performance of glass varies according to the type of earthquake experienced. When detailed properly with enough tolerance around the frames in order to allow for movement, glass has performed quite well in accommodating long-period motions. Contrary to popular thought, when the glazing system is well detailed to allow for movement, all the glass of a building does not cascade into the streets below. However, since it is impossible to predict the type of earthquake and its exact motions that will occur at a given time, the hazard posed by an element such as brittle glass will always exist. There have been enough examples of glass failures, with shards of glass falling into the street, that the hazard has to be considered and mitigated in seismic building design. (See Figure 6-6.)

Full-Scale Laboratory Tests: Exterior Cladding of a Six-Story Steel Moment-Frame Building

In 1984, as part of the U.S./Japan Cooperative Research Program, laboratory testing of a full scale six-story steel moment-frame building with exterior cladding applied took place in Tsukuba, Japan, under the auspices of the Building Research Institute. The cladding materials used were precast concrete and glass-fiber reinforced concrete (GRFC) panels that are commonly used on the West Coast of the United States. Professor Marcy Wang, from the Department of Architecture of the University of California at Berkeley, was the principal investigator (Wang, 1987).

The test loading sequence culminated in a 1/40 story drift ratio, which closely corresponds to both a drift postulated in a major credible earthquake and to UBC design standards for drift requirements. Two types of connections were utilized to attach the nonstructural cladding elements to the structural frame: (1) a "sway" or "translation," mechanism commonly used in the United States; and (2) a "rocking" mechanism commonly used in Japan. Some examples of representative test results indicate that

1. Long ductile rods used for lateral connections can accommodate large story drift. However, sliding connections may have problems due to insufficient slot length or potential impedance of the sliding mechanism due to weathering, aging, improper installation, or poor detailing.
2. Slot lengths need to be generous to avoid imposition of unexpected large stresses in the connections and on the panels themselves.
3. Bearing connections must be sufficiently flexible to avoid conveying stress, resulting from interstory drift, directly to the panel in both in-plane and out-of-plane directions.
4. Care must be taken not to stiffen connections inadvertently; for example, spilling new concrete around the connecting elements.
5. Panels should be "hung" by detailing bearing connections at the top and lateral connections at the bottom. When this detail is reversed, as is common practice in some areas, so that the bearing connection is at the bottom, the panels may rotate, overturn, and fall out if the lateral connections fail at the top.
6. Joints must have ample tolerance to avoid contact between panels as they accommodate for drift, or crushing and failure of panels will occur. The design of cladding for corner conditions is particularly important to avoid contrasting and compounded movements.

LIFE SAFETY ISSUES

The principal issue governing the mitigation of seismic hazards ascribable to nonstructural performance failure and/or damage is concern for public health and safety. Life hazard as related

Figure 6-7 Overturned metal library stacks, 1972 Managua, Nicaragua, earthquake. *Source:* Henry J. Degenkolb.

to nonstructural elements is generally categorized into three components:

1. *Primary elements,* such as exterior cladding/veneers, infill walls, rigid stairways, parapets/canopies, and so on.
2. *Secondary elements,* such as glazing, partitions, ceilings, elevators, equipment, shelving, light fixtures, contents, and so forth.
3. *Supply system elements,* such as life-support systems (hospitals), toxic chemical containers, electrical equipment, and so on.

A representative sampling of life safety exposure to hazardous nonstructural damage in past earthquakes includes: (1) collapse of suspended light fixtures; (2) falling shards of shattered glass; (3) falling exterior cladding materials; (4) overturning of heavy equipment; (5) falling parapets; (6) collapse of hung ceilings; (7) overturning of book shelves, storage racks, and library stacks; (8) collapse of infill walls; (9) falling wall-mounted ornamentation/artwork; and (10) rupture of toxic chemical containers. (See Figures 6-7 and 6-8.)

Performance of Nonstructural Elements and the Time of Day of Earthquakes

Levels of life safety exposure to an earthquake are directly related to three fundamental variables: (1) the size of the earthquake, (2)

Figure 6-8 Debris from suspended ceiling, HVAC ducts, and light fixtures after 1972 Managua, Nicaragua, earthquake. *Source:* Henry J. Degenkolb.

the performance of the building system and nonstructural elements, and (3) the number of occupants in the building at the time of day when the earthquake occurs. Obviously, if no one is in the building at the time, no injuries or life loss will result even with severe nonstructural damage or a "pancake" structural collapse.

In contrast to the late evening hours, when most people are at home, where smaller rooms hold fewer occupants, during normal working hours, when public schools, commercial buildings, and public offices are fully occupied, exposure of life safety to failure of nonstructural elements increases exponentially. Accordingly, statistics on life loss and injuries due to nonstructural elements are somewhat skewed. Depending on whether a major, damaging earthquake in a central downtown area occurred at around 2 to 3 P.M. during a weekday or at about 9 to 10 A.M. on a weekend, morbidity statistics could vary dramatically.

Because of this, seismic hazard exposure to life posed by the nonstructural elements of a building system is perceived on different scales by many design professionals. Indications are that our views on the subject depend on personal experience and imagination. There is much hard data available on which specific nonstructural elements failed during an earthquake, but little quantitative material on specific deaths and injuries actually resulting from failures of specifically identified nonstructural elements. Accordingly, all we have left at the moment, in general, are accounts such as mentioned earlier in the 1964 Alaska earthquake: "At least three deaths, numerous injuries, and considerable panic were caused by the failure of nonstructural elements." Although many researchers are currently at work on the subject, a truly definitive measure of the manner in which a nonstructural element is directly responsible to life loss and injury (or merely an inconvenience) is generally not available, except for some data collected after the 1964 Alaska, the 1985 Mexico, the 1988 Whittier-Narrows, and the 1988 Armenia earthquakes.

One part of the answer may come when final research results on the 1985 Mexico earthquake become available. There was a great deal of nonstructural damage documented in newer buildings located in Mexico City, aside from those that collapsed. Data on the subject has been collected, with final reports issued in late 1989 and early 1990. Another part of the answer will come when a major earthquake impacts a metropolitan center in the United States, testing the performance of nonstructural elements in newer buildings constructed under U.S. codes, standards, and construction methods. The October 1989 Loma Prieta earthquake in California provided additional information on this subject (see Chapter 12).

FUNCTIONAL IMPAIRMENT OF BUILDINGS

In addition to concerns for life safety, damage to a building's nonstructural elements may also lead to its inability to continue to function as intended; likewise, its serviceability may no longer be viable. With failure of nonstructural components on a major scale, as recorded in the 1985 Mexico City earthquake, attention has focused on elements that when severely damaged lead to building outages and functional impairment.

In addition to life safety and community serviceability, there are economic concerns. The potential of having several buildings in an investment portfolio functionally impaired and out of service for a long period of time, with extended loss of income, owing to major nonstructural damage is also unacceptable to clients. This situation would be particularly embarrassing for the architect in situations where the building's basic structure may have performed well seismically. Picture yourself phoning a client after an earthquake to say, "Congratulations, your earthquake-resistant structural system survived the earthquake most successfully and it is still standing tall!! But, it will cost you about 65 percent of the building's replacement cost before it can be reoccupied. Furthermore, it will be out of service for a minimum of six months while we put all the nonstructural elements back together again."

This aspect of the problem is especially important when we are dealing with critical emergency service facilities such as acute hospitals, police stations, fire stations, airports, power plants, and water supply and treatment plants, among others. Without the services provided by such facilities, a community today would not survive very long.

Representative Examples of Functional Impairment

Even if structural failure and/or building collapse does not occur, a fire station may still not be able to move out fire-fighting equipment to respond to community needs if its main exit doors are jammed shut by an earthquake. In one such case in the 1980 Irpinia, Italy, earthquake, jammed metal doors had to be cut open with an acetylene torch before anything could be moved out. By the time that was done, a fire had flashed out of control.

A financial facility, such as a bank, must remain operational on an active basis to provide the community with continued economic support and other essential financial transactions during postearthquake recovery efforts. If its data-processing capacity is down following an earthquake, loss of function will result. Following the October 1989 Loma Prieta earthquake in California, several major banks in San Francisco found that their ATM machines were out of order for as long as 24 to 36 hours.

An acute hospital should remain functionally operational after an earthquake in order to continue to serve its inpatients as well as maintain its position to treat earthquake victims. Obviously, a major hospital building should be a serviceable asset after a damaging earthquake rather than become a liability during recovery efforts and add to the confusion. Everyone seems to agree on this. Yet, oddly enough, after the 1971 San Fernando earthquake, the reverse actually occurred. (See Figure 6-9.)

Figure 6-9 Ambulances crushed by collapse of reinforced concrete roof, Olive View Hospital, after 1971 earthquake in San Fernando, California.

Special California Earthquake Legislation

The 1971 San Fernando earthquake was particularly destructive to medical facilities located in the area. Their impairment increased the inability of hospitals to offer emergency care services in treating the injured, let alone take care of their own patients, many of whom had to be evacuated from the premises. Of the four major hospitals located in the San Fernando region, one collapsed and the other three were evacuated owing to significant damage. Of the three medical office buildings, two psychiatric units and one mechanical equipment hospital service unit in the area, all except one (a psychiatric unit) were damaged significantly.

As an aftermath of the San Fernando earthquake, it was most clear that health care facilities are a type of building especially important to immediate postearthquake recovery efforts. In this case, their damage and impairment were particularly disruptive to the treatment of injured, and the hospitals became liabilities to recovery activities. In analyzing the impact of the San Fernando earthquake, Kesler (1973b) writes:

> Not only are patients incapacitated in many cases and unable to take, perhaps, even simple precautions to protect themselves, let alone safely endure an interruption in care, but also medical facilities are urgently needed in the hours following widespread destruction and injury due to an earthquake. At the time of disaster, their installations must be functional, rather than being among the casualties.

Public and political reaction to the questionable performance of the four major hospitals in the San Fernando area resulted in legislation designated as the "Hospital Safety Act of 1972," which became effective in March 1983. In passing the Hospital Act, it was the specific intention and mandate of the California legislature that any new hospitals constructed in California "be completely functional to perform all necessary services to the public after a disaster."

The end result of this legislation was a new concept in seismic damage control in that not only the structural system, but also all critical nonstructural elements (including architectural, mechanical, electrical, and life support systems), are expected to maintain their integrity and remain operational after a major, damaging earthquake. Now, for the first time, we have nonstructural elements gaining a distinct prominence in requirements related to earthquake hazards reduction. The intent of the legislation is not only that a hospital remain "undamaged," but that it remain free from critical nonstructural damage as well. The requirements for damage control specify that critical nonstructural building elements, systems, equipment, and apparatus necessary for the complete function and operation of a hospital are to be "designed, detailed, and constructed to withstand the maximum acceleration and deflections of the basic structure without excessive displacement or damage that would disrupt essential operations and services to be performed." Chapter 10 also covers this subject as part of earthquake hazards reduction goals.

Damage Control Standards

Performance standards for damage control are typically determined at levels that permit a building to continue to function. The intent of these standards normally is not set to preclude damage in general, but rather limit specific damage patterns that result in serious building impairment. For example, in the extreme case of a hospital in California, standards give particular attention to the deflection of wall assemblies, glazed openings, and anchorage details, among other requirements. The horizontal deflection of vertical structural systems due to lateral forces in the plane of the wall is not to exceed 1/16th of an inch per foot of height of any story. Deflection from head to sill of glazed openings, in the plane of the wall, is not to exceed 1/32nd of an inch per foot of height of the opening unless the glass therein is prevented from taking shear or distortion, or where tempered, safety or wire glass is used.

To indicate the detail under which nonstructural aspects of a building are covered by California's 1972 Hospital Act, the anchorage of all fixed items and components is also included. The bracing and/or fastening of major equipment and critical movable apparatus, such as autoclaves, sterilizers, kitchen fixtures and appliances, laboratory materials, X-ray equipment, and cubicle enclosures, must be detailed in consultation with the engineer, or seismic technical consultant, of record. In the architectural set of drawings, the manner in which all nonstructural partitions, window–wall assemblies, and wall openings are attached, fastened, or connected to other components and systems of the structure must be completely accounted for with construction details.

Damage control requirements have a specific impact on the design of architectural, mechanical, and electrical components when their potential failure may be serious enough to put the building out of service for a long period. Each element, so identified, must be examined in detail to determine its role and its importance factor in maintaining the continued function and operation of the facility.

Design solutions of nonstructural elements should be assessed to take into account (1) deformation compatibility of structural and nonstructural components, and (2) inelastic deformation of any portion of a connection between components that could create functional impairments.

How Far We Should Go, Not How Far We Can Go

The danger of damage control standards is that once we start on the path of striving to keep buildings functional, it is difficult to know when to stop. Because it is very easy to go overboard, it is necessary to keep things in proper perspective by maintaining objectivity and remembering that we are not designing "earthquake-proof bunkers," but rather typical building types in a community environment with assigned cost constraints.

Consideration of total construction costs will provide the ultimate overriding control, but philosophically it is important to ask key questions: How much damage control is acceptable? Can damage be totally avoided? What level of design effort is prudent to achieve appropriate safety? Is it possible to design an "earthquake-proof" building, and if so, is it cost-effective to do so? Care must be maintained to preclude overly zealous approaches to this problem, as efforts greater than necessary may be expended to solve a minutiae of problems not critical to life safety or damage control objectives.

Obsessed by liability issues, besieged by the need to improve marketing strategies, and concerned about profitability, architects already may feel that they have enough to worry about without the added burden of seismic design of nonstructural elements. Yet, when competing for contracts in regions of high seismic risk in locations throughout the world, if the architect wishes to maintain a prominent position in the marketplace, seismic issues must be addressed. The 1985 Mexico earthquake proved this point in Mexico City, where architects, rather than structural engineers, found themselves accused of being negligent and ignorant of earthquake-resistant design principles when their buildings had to be evacuated owing to excessive damage.

BUILDING ACCESS/EGRESS

Requirements for safe exiting from a building, particularly from a major structure such as a high-rise office or hotel building, must include adequate, unobstructed circulation paths for egress, emergency lighting, and freedom from falling debris. Today it is not unusual to find many of these "superscraper" high-rises with a population of 10,000 occupants or more. To make the picture more complex, it is also important to remember that immediately following a major earthquake, electric power and telephone communications systems have not been available. At times like this, one only hopes that the emergency power generator, a nonstructural element, was bolted down, remains undamaged, and functions properly.

Conversely, when emergency situations arise search and rescue teams must be able to enter a building freely and immediately after a damaging earthquake, encounter minimum interference, blockages, and damaged nonstructural elements in corridors, and take care of seriously injured occupants and locate those trapped by fallen debris. They need to find control and emergency communications substations operational to cope with potential fires, toxic spills, and other hazards that might occur after the earthquake. The teams also want to remove the immobilized injured from the damaged, weakened building before another serious aftershock occurs.

Common Problems Impeding Access/Egress

Examples of nonstructural damage impairment problems impeding access to and egress from multistory buildings documented in past earthquakes (Figures 6-10 through 6-13) include the following, some of which have been purposely cast as worst-case scenarios:

1. Stairways and corridors not negotiable, blocked by debris from wall enclosures and ceiling materials, compounded by electric power outages and unavailability of emergency lighting, communications systems, ventilation, and so on.
2. Exit doors and/or fire doors jammed by deformed door jambs and have to be pried open (or, in worse cases, torched open), and so forth.

Figure 6-10 Collapsed stairway and shattered infill walls, 1985 Mexico City earthquake.

Figure 6-11 Shattered curtain wall and collapsed reinforced concrete stairway, 1985 Mexico City earthquake.

3. Office doors blocked by heavy furniture, file cabinets, and/or other equipment that has overturned or shifted location by "dancing" across floor, and so on.
4. Elevators rendered inoperable owing to loss of electric power, and/or twisted cables, swinging counterweights and damaged controls, and so forth.

BUILDING CONTENTS

A great deal of attention is currently being given to large economic investments represented by the contents of buildings: (1) high-tech products of light industrial electronic manufacturing buildings, (2) financial records in banking institutions with a high utilization of data-processing equipment, and (3) critical medical facilities such as blood banks and others. In addition, concerns are also being ex-

Figure 6-12 A collapsed elevator penthouse that rendered elevators inoperable. *Source:* K. V. Steinbrugge private collection. Reprinted with permission.

Figure 6-13 Impassable debris-laden exit stairway, 1972 Managua, Nicaragua, earthquake. *Source:* Meehan et al. (1973). Reprinted with permission.

Figure 6-14 Stone stela braced with metal brackets at sides and bottom, Museum Of Anthropology, Mexico City.

pressed for the storage of toxic, chemical, explosive, and radioactive materials in many laboratories and manufacturing plants.

The value of building contents will, of course, vary greatly depending on the usage of the building, which may range from a simple office building to a sophisticated high-tech operation. The greatest costs are associated with high-tech buildings, such as data-processing centers, chemical storage facilities, research laboratories, communications centers, major hospitals, financial records institutions, and electronic component manufacturers. Recent additions to the list now include major museums which may contain priceless "one-of-a-kind" objects either on display or in their archives.

After the 1985 Mexico earthquake, everyone breathed a sigh of relief when it was known that the priceless contents of the Museum of Anthropology in Mexico City suffered little or no damage. Was this a miracle within itself? No, it was not, but rather it was a result of the fact that the museum conservators were cognizant of the need to protect the collection from seismic effects. Figures 6-14 and 6-15 illustrate methods that were used to protect the priceless pieces on display.

Figure 6-15 Circular relief stone braced with metal saddle, Museum of Anthropology, Mexico City.

Long-period motions have more of a slow swaying effect compared to short-period motions, which result in sharp, jolting accelerations more capable of jarring objects off shelving and from their protective mountings. If the earthquake had been of shallow depth located immediately under Mexico City, the effects of ground shaking and accelerations on such building contents would have been distinctly the reverse of what happened in 1985.

Economics of the Building Contents Problem

Since it is known that damage to building contents may result in severe economic loss, the importance of their protection and safeguard should not be underrated. It has been estimated that economic loss from all aspects of nonstructural failure could easily run as high as 10 times the construction value of the building if one considers loss of architectural components, expensive equipment, inventory, use of the facility until it can regain operational capability, staff employment, and projected sales income. The point that needs to be emphasized is, even though many design professionals and the media tend to stress structural damage in earthquakes, in certain situations the cost of damage to nonstructural elements may exceed that of structural elements. Moreover, costly damage to nonstructural elements can occur in earthquakes of moderate intensities that may cause little or no structural damage.

NONSTRUCTURAL DAMAGE MITIGATION METHODS AND PROCEDURES

Movable Components and Equipment

There are many options and strategies available for the reduction of nonstructural damage and the protection of building contents, depending on the features and importance of the components and equipment involved. In general, mitigation techniques for protect-

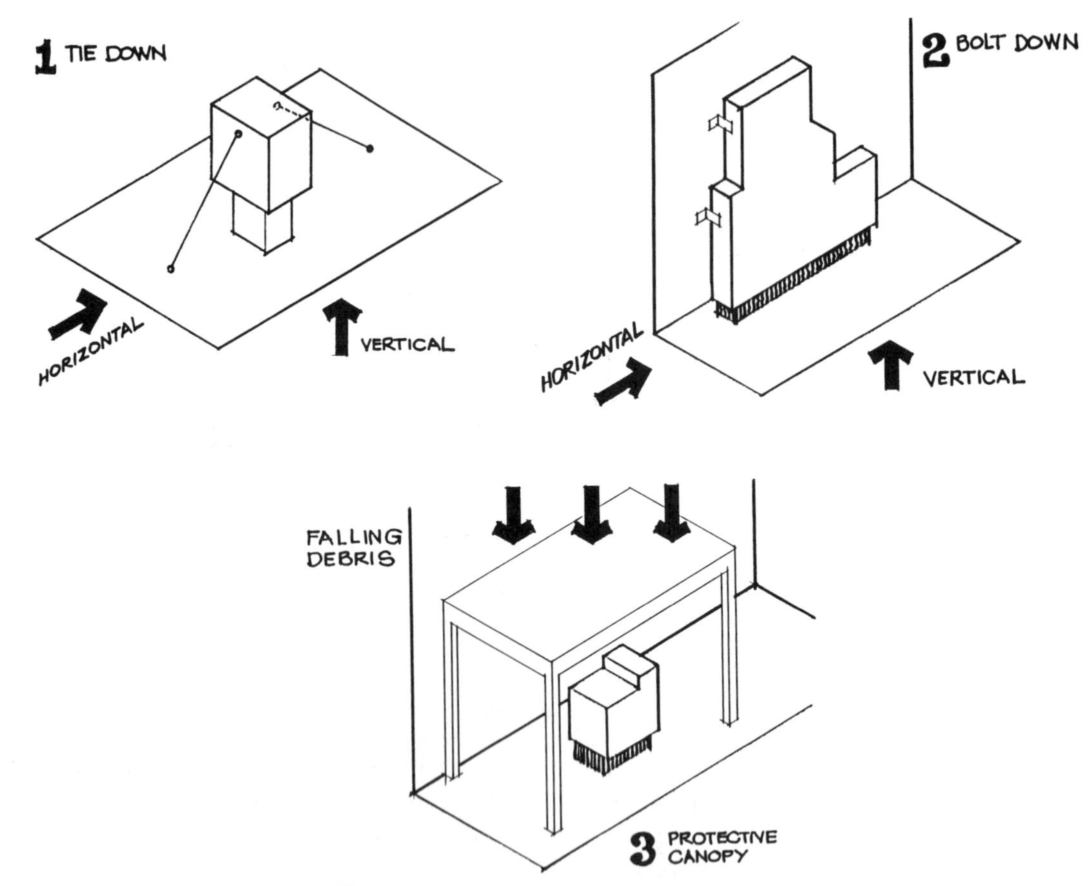

Figure 6-16 Building contents and equipment prototype hazards mitigation techniques.

Figure 6-17 Removable metal shelf guards for critical emergency supplies, Tokyo Emergency Response and Services Office, Tokyo, Japan.

ing movable building elements include: (1) bolting down, (2) tying down, (3) bracing, (4) using protective canopies, and (5) employing base isolation. These techniques also apply to mechanical and electrical components and equipment. (See Figure 6-16 for general representations of sample mitigation techniques.) The placement of removable shelf guards is an appropriate solution when dealing with storage of critical materials such as toxic chemicals and medical supplies (See Figure 6-17).

Fixed-In-Place Elements

When dealing with nonstructural elements to be "fixed in place," such as interior partitions, infill walls, stairways, or other such "built-in" building components, another approach must be used. It must first be decided whether the element should be: (1) relatively detached from the basic structure with slip-joints and/or other appropriately flexible connections to allow for relatively free structural movement without damaging or impacting the nonstructural element or the structural member itself; or (2) firmly attached to the basic structure if it is possible to rely on the nonstructural element to withstand building movements, shaking, and induced stresses without critical failure.

One purpose for using the first option is also to avoid having the nonstructural member assume structural characteristics during an earthquake and thus negatively change the intended design performance of the basic structure itself. In the second option, the nonstructural element is assumed to have the capacity to move right along with the structure itself without crippling damage.

Recommended approaches may also depend on how the design professional visualizes and anticipates the overall seismic performance of the entire building system as a whole. One must project whether the anticipated building system will be relatively rigid and projected movements small, or whether it will be relatively flexible and the expected movements large. The approaches also depend on the characteristics of the materials of the nonstructural elements themselves; that is, whether they are brittle or ductile. For example, for nonductile materials that do not have the capacity to deform easily without failing, such as glass, brittle elements must be appropriately detailed and set in place with tolerances for excessive movements to avoid shattering.

Besides the traditional methods for reducing nonstructural damage described above, recently there has been increased attention paid to decreasing earthquake effects on buildings or their elements by installing seismic motion isolation and/or energy absorption devices. Although earthquake motions and deformations induced in a building can be greatly reduced by these isolation systems, they cannot be totally eliminated, so the strategies indicated above must always be considered in building design. On this basis, the assumption is that seismic motions, or equivalent static forces, can be determined and appropriate protective measures developed.

Special Building Contents and Equipment

With the recent advent of "smart" buildings, the growing importance of communications systems, and the increased reliance on computer centers as electronic data-processing facilities, contents and equipment have become a special, separate nonstructural category. Distinct hazards, which can cause critical damage or downtime to building contents and/or equipment, are divided into four areas:

1. *Overturning* (contents/equipment falling over),
2. *Sliding/rotating* (sliding, rotating, or swinging of mounted/hung contents/equipment),
3. *External debris* (danger of damaging debris cascading from ceilings or walls on contents/equipment), and
4. *Loss of utility services.* (It is a common experience during major earthquakes to expect loss of electric power and communications lines. Without power, air-conditioning systems would also be functionally impaired.)

As indicated by mitigation methods indicated earlier in this chapter, development of programs for the seismic protection of contents and equipment, including data-processing facilities, requires the design of appropriate anchoring, restraining, or isolation devices as well as outright protection from debris falling from the ceilings.

Raised computer floors are vulnerable to failure during extreme shaking. It must be constantly kept in mind that earthquake loads induce horizontal forces. These forces automatically induce horizontal deflections in the raised floors, which can lead to failure. In addition, vertical earthquake motions should also be considered. Normally these effects are neglected under conventional building code practice. However, the use of flexible framing to support floors may amplify vertical earthquake motions acting on the floor. In short, to avoid overturning and/or shifting of computer equip-

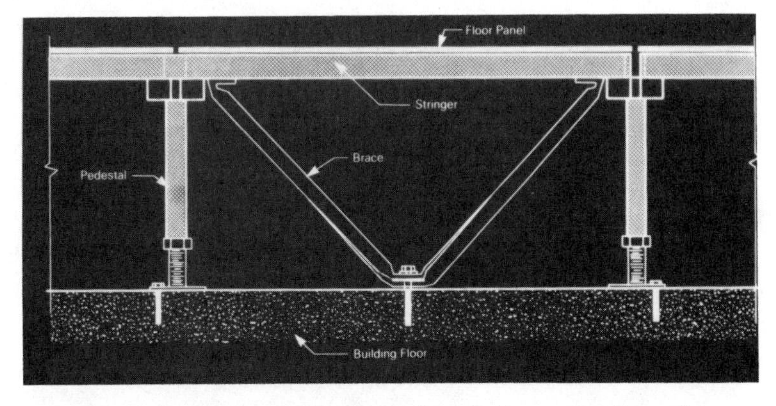

Figure 6-18 Prototype bracing system for raised computer floors. *Source:* Olson et al. (1987). Reprinted with permission.

Figure 6-19 High-tech communications system console braced with metal channels to wall.

ment, (1) vertical supports for raised computer floors must be braced, and (2) floor panels must have a tie-down system installed.

Computer equipment itself must be tied down, bolted, and/or braced to mitigate damage. Care must be taken not to stop such connections at the floor panel, which easily pull out and may thereby negate the whole process; rather, the system must be connected to elements capable of restraint, such as the structural floor or continuous structural stringer.

Finally, it would be useless to protect computer equipment and appropriately brace raised-floor installations if mitigation techniques were not extended to protect tape and disk storage, the loss of which could be critical as well. First, the tapes must be restrained from rolling off shelves or the storage unit prevented from overturning.

In short, any method selected to reduce damage to contents and equipment, including data-processing facilities, must be thoroughly thought through in a systematic way to be effective. It must not stop at the "weak link" for the earthquake forces to find and dismantle. (See Figures 6-18 and 6-19.)

SUMMARY

Earthquakes can cause extensive damage to the nonstructural elements of a building system as well as the building's contents, not only to the basic structure. In many cases it has been calculated that the cost of repairing nonstructural damage can be considerably greater than the cost of structural repairs. If appropriate mitigation measures are taken into account, most of the damage to nonstructural elements and building contents can be avoided at little or no extra cost when included in the original design of the building.

CHAPTER 7

EXISTING BUILDINGS

In previous chapters the primary focus was on the seismic design of new buildings. With this chapter, however, the emphasis will be shifted onto a major problem faced by urban centers located in earthquake-vulnerable areas: what to do about the seismic performance of all those existing buildings that were designed under code standards in force at the time of their construction. Although still serviceable, many of these existing, older buildings found in the downtown areas of our cities, or in their central business districts, would be categorized as at earthquake risk simply for not meeting seismic requirements specified by the latest edition of the building code.

Since building codes are not retroactive, these existing buildings, though legally acceptable, may be extremely hazardous from a seismic point of view. Depending on their type of construction and the code provisions under which they were originally designed, it is known that many of these existing building classes have a poor performance record relative to earthquakes.

Fortunately, technical guidelines have been developed by which the relative hazard posed may be assessed, analyzed, and potentially mitigated. However, although we are successfully approaching technical solutions to the problem, we have yet to solve the social and economic aspects of the dilemma presented by the phasing out of these existing seismically hazardous structures, which constitute a vital part of our urban building stock available for housing and commercial purposes.

HISTORIC PERSPECTIVE

Historically, the traditional approach to improving the performance of structures and upgrading building code provisions has focused on the advancement of design standards for new buildings. Over the years, seismic requirements in building codes have placed greater emphasis on new building design than deriving standards for existing buildings. During this period, accordingly, very little, if any, code provisions are found with reference to the seismic upgrading of older, existing structures except in the context of "bringing it up to full code" when triggered by additions and alterations that exceed a specified limit in area and/or dollar value of the building.

Recalling that the first seismic code in the United States was established after the 1933 Long Beach earthquake, it is easy to understand why the seismic performance of pre-1934 buildings would be highly suspect. (See Figure 7-1.) As no seismic prerequisites were required in any building constructed prior to 1934, it could only be concluded, as a general rule, that structures in this category were not designed to earthquake-resistant standards, and accordingly, their seismic performance would be marginal at best during a substantial, damaging earthquake. Over 50,000 of pre-1934 unreinforced masonry bearing wall (URM) buildings still exist in California.

It was not until 1978 that a major attempt was made to deal with the seismic problem posed by older, existing hazardous buildings on a comprehensive basis. For building code purposes, performance standards on this problem were first developed and presented by the Applied Technology Council of the Structural Engineers Association of California (SEAOC). The problem associated with upgrading the seismic performance of older, existing structures was identified as a critical one to be addressed seriously by design professionals, researchers, and public policy officials.

Over the years, further concerns about the seismic performance of older buildings arose when it was realized that, by far, the majority of structures in our metropolitan centers were those older buildings, still in use though designed under previous code standards. Technically, as explained earlier, the strictest assessment of such buildings would result in their potentially being judged hazardous when mea-

Figure 7-1 Severely damaged and partially collapsed unreinforced masonry building, 1983 Coalinga, California, earthquake.

sured against current code provisions. In contrast to the percentage of older, existing buildings found in the urban centers, recent surveys reveal that on an annual basis nationwide, the construction of new buildings only accounts for an addition of about 2 percent per year to our total building stock. The implication here is that the remaining buildings completed in prior years represent 98 percent of the building inventory of our metropolitan centers.

PLANNING AND DESIGN PROBLEMS

Planning and design problems to be addressed in consideration of existing hazardous buildings may be measured against the assessment of three primary concerns:

1. Life loss and injuries,
2. Direct property damage, and
3. Functional impairments.

The characteristics of life loss and injuries are functions of: (1) the building's construction type and number of occupants, and (2) off-site geological effects having consequential performance impacts. A relationship exists between buildings and casualties that is directly aligned to building classes and construction types and their anticipated seismic performance. Past studies have plainly demonstrated that certain construction types such as single-story, lightweight, wood-frame structures (without excessive heavy roof or floor loads) will survive threats of total "pancake" collapse quite well. (See Figure 7-2.) Conversely, unreinforced masonry with poor mortar is usually associated with heavy life loss and property dam-

Figure 7-2 Damaged lightweight wood-frame residence, knocked off foundations but not collapsed, after 1983 Coalinga, California, earthquake.

age contingent on the type of acceleration and period of ground motion experienced.

RELATIVE EARTHQUAKE SAFETY IN BUILDINGS

Most current seismic building codes are intended to protect life and reduce (not eliminate) property damage. Even though we cannot now predict earthquakes with respect to specific time, location, and magnitude, it is apparent from past experience that existing buildings located in regions of high seismic risk will be exposed to a major earthquake at one time or another. The potential of death or injury to people living or working in potentially hazardous building types is clearly identified as a significant social concern.

By establishing 1934 as an applicable base, prior to the promul-

TABLE 7-1 Relative Earthquake Safety of Buildings

	Probable Life Loss Per 1000 Occupants		
Building Type	Earthquake-Resistant Building		Non-Earthquake-Resistant Building
Small wood frame	2		4
Large wood frame	5		10
Small, all metal	2		4
Large, all metal	8		15
Steel frame, superior	5		10
Steel frame, intermediate	10		25
Steel frame, ordinary	15		40
Steel frame, mixed with wood floors	25		50
	Ductile Concrete	Nonductile Concrete	
Reinforced concrete, superior	25	50	100
Reinforced concrete, intermediate	50	200	500
Reinforced concrete, ordinary	75	300	1000
Reinforced concrete, precast	75	500	1500
Reinforced concrete with wood floors	100	800	2000
Mixed construction, superior	15		800
Mixed construction, intermediate	20		1000
Mixed construction, ordinary	40		2000
Mixed construction, URM			4000
Mixed construction, adobe/hollow tile			5000

Source: Steinbrugge et al. (1979).

gation of seismic codes and the development of earthquake-resistive design, it is possible to derive the extent of maximum probable deaths in representative building classes. In doing so, it is necessary to exclude external geological effects on structures, such as landslides, liquefaction, subsidence, flooding, or tsunami.

By focusing on a building's anticipated seismic performance alone, without the consequential effects of any geological impacts as mentioned above, an analysis of the maximum probable deaths per building type per 1000 occupants in a typical urban center located in a high-risk area can be developed. Table 7-1 indicates the results of this analysis as a function of the relative earthquake safety of buildings.

It is important to realize that the figures in Table 7-1 indicate general projections of life loss due to the relative safety of typical structural systems generally found in the existing building stock and do not include collateral seismic impacts, such as soil failures and other geological effects. In this regard the figures are useful as a general measure in identifying the relative safety of representative building types in the form of a life safety ratio, (see Steinbrugge, K.V., and Lagorio, H.J., 1985).

Public Policy and Public Safety

From a policy point of view toward public safety, an initial goal would be first to strive to reduce the risk of life loss to 2 per 10,000 persons for all types of buildings in the event of a maximum credible earthquake. This recognizes that the current state of the art of earthquake engineering does not allow for "earthquake-proof" buildings in terms of the expected cost benefit returns, and that the eventual limiting value is currently estimated at a 2:10,000 ratio. The 2:10,000 figure is based on studies of past damaging earthquakes and represents the lowest expected loss of life in small, light wood-frame structures, which, because of their flexibility, redundancy, and light weight are considered to be the safest type of structure in general use in the United States. As previously indicated, it is very rare to see a total "pancake" type of structural collapse in this class of building. (See Figure 7-2.)

In summary, in analyzing the relative seismic safety of existing building classes, recognition is given to four fundamental considerations:

1. Current state-of-the-art practice in earthquake engineering does not anticipate producing "earthquake-proof" buildings in terms of cost-effective construction systems.
2. Projected lower bound limits for life loss are estimated at the 2:10,000 ratio due to the positive seismic performance of small wood-frame buildings of light construction.
3. Projected higher bound potential limits for life loss are esti-

mated at the 4000:10,000 ratio for pre-1934 URM structures with occupancies and uses represented by older commercial and manufacturing building types.

4. A building's age, size, type of construction, and quality of maintenance have a direct bearing on its expected seismic performance and relative safety.

SEISMICALLY VULNERABLE TYPES OF CONSTRUCTION

After each major earthquake new lessons are learned, or old ones relearned, through field investigation concerning the performance of specific building types. These lessons, including those derived from the 1985 Mexico and the October 1989 Loma Prieta, California, earthquakes, result in the advancement of new seismic standards in building code performance requirements. Accordingly, with progressive adoption of new seismic code standards based on new lessons learned, a greater number of existing structures face the possibility of being identified as technically hazardous.

The anticipated seismic performance of a building system is therefore generally classified according to its date of construction (building age), building class, construction type, and correlation with the year of adoption of seismic code standards currently in use. For example, earthquake-resistant reinforced masonry design was an unknown type of construction before the 1933 Long Beach, California, earthquake. It was the URM building type that was severely damaged, with many collapses in evidence, at the time of the 1933 earthquake.

In the same way, subsequent code changes, or major additions to code provisions, will impact the perceived seismic vulnerability of older buildings when compared to the levels of performance and construction practice in effect when the structures were originally designed. Obviously, any such deficiencies would vary widely, depending again on: (1) the age of the building and adequacy of code provisions at the time of construction, (2) building type and class, (3) design and construction practices, and (4) building size and configuration.

Seismic Risk Analysis and Existing Buildings

The seismic risk of a single building is generally defined as a combination of four factors: (1) hazard, (2) exposure, (3) vulnerability, and (4) location. Hazard includes all possible geological hazards such as strong ground-shaking potential, fault rupture, liquefaction, and landslides, among others. Exposure refers to the peril in which public health and safety is placed in the face of the hazard, and also includes the occupancy and function of a building.

Vulnerability is associated with the expected performance of the building system. Location is the proximity of the building to a potential earthquake source. To identify high-risk buildings, all four factors should be considered. Thus, a building of a potentially vulnerable construction type may not be considered a high risk if it is located in an area not even remotely exposed to an earthquake source. Nevertheless, the establishment of characteristics identified with potentially vulnerable building classes of high risk is necessary as a first step toward developing an earthquake hazards reduction program.

Even vulnerable structures do not necessarily, or automatically, produce a high life-loss seismic risk, since life loss is also associated with the use, or occupancy, of the building. Obviously, a building filled with many occupants poses a greater risk to life loss than an empty building. Hence, it is not the intent of this section to imply that all buildings identified as potentially hazardous construction types are extremely vulnerable structures, or to conclude that all buildings included in this category are of high risk; rather, it is to emphasize that many variables are present in these buildings that may render their anticipated seismic performance potentially hazardous. As a consequence, they merit detailed examination and analysis prior to coming to any conclusions relative to earthquake safety.

Potentially High-Risk Buildings

Architects must realize that while the unreinforced masonry (URM) bearing wall building has gained much notoriety, other specific construction types have been assessed as presenting a potentially equal or even greater hazard to public safety. One of the other such types most frequently cited is the nonductile reinforced-concrete-frame-type building constructed from the early 1940s to the early 1970s, before the development of ductile concrete theory. As another example, since the advent of precast, prestressed reinforced concrete systems in the late 1960s, deficiencies in the early types of this building system have appeared, also in contrast to currently defined earthquake-resistant design standards.

As a result of recent research activities by the earthquake engineering community, including the California Seismic Safety Commission (CSSC), the Earthquake Engineering Research Institute (EERI), the Applied Technology Council (ATC), and the Center for Environmental Design Research (CEDR), a minimum of seven classes of older, existing hazardous building types of construction and more recent building classes have been identified as being potentially dangerous under earthquake loads. These seven (7) classes are presented in Table 7-2.

Building diagrams shown in Figures 7-3 through 7-11 illustrate the typical construction elements and general characteristics of some categories of potentially hazardous buildings listed in

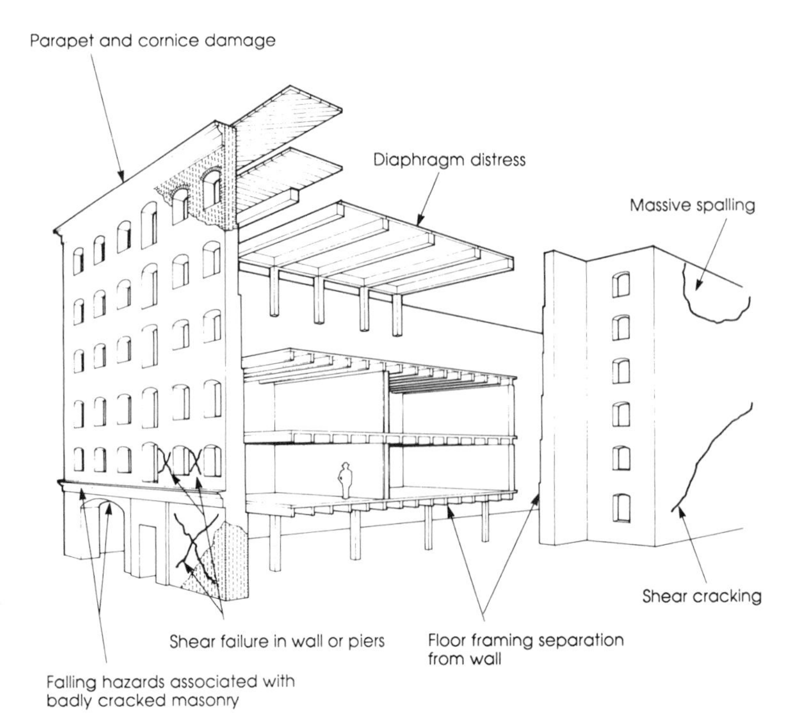

Figure 7-3 Unreinforced masonry bearing wall building. *Source:* Rojahn and Reitherman (1989), ATC-20, Applied Technology Council.

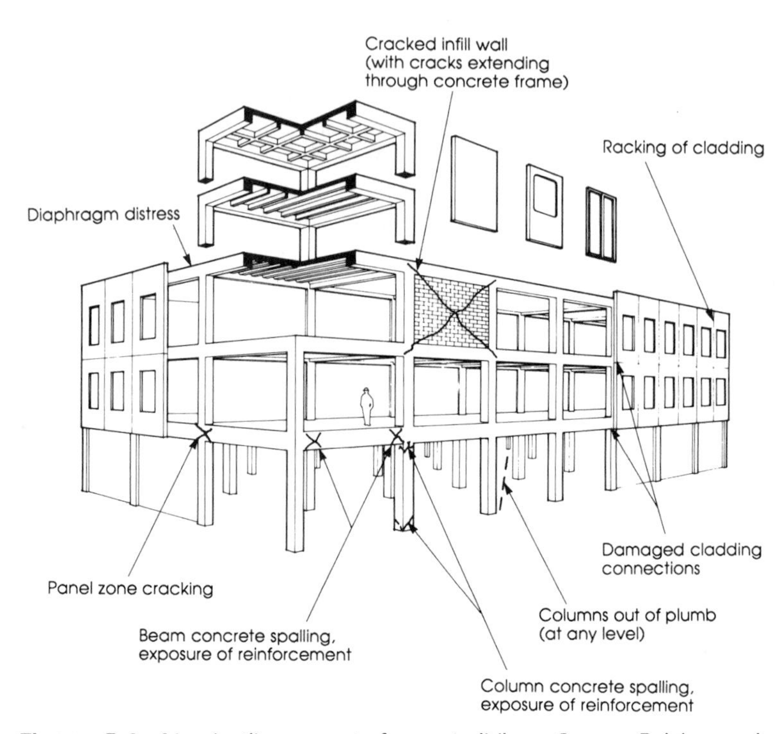

Figure 7-4 Nonductile concrete-frame building. *Source:* Rojahn and Reitherman (1989), ATC-20, Applied Technology Council.

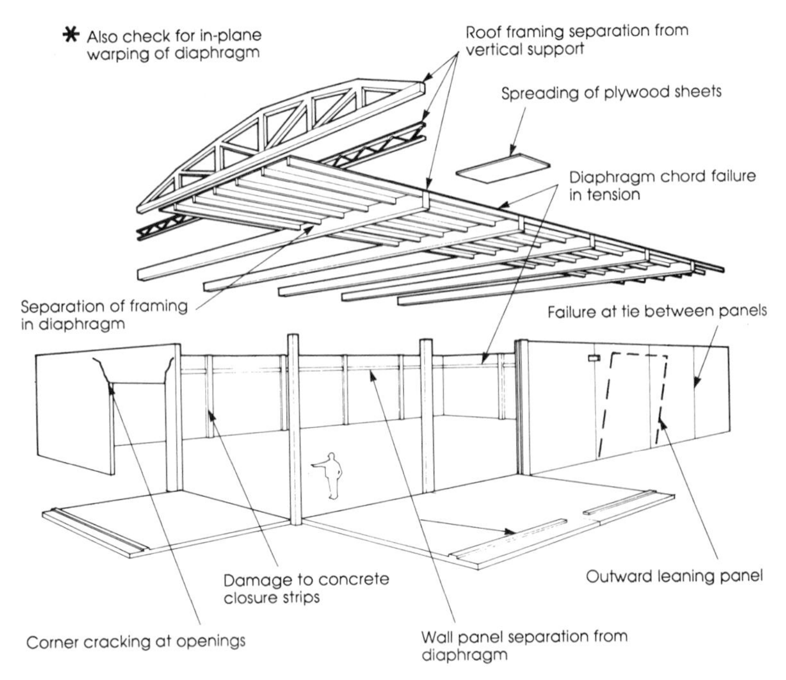

Figure 7-5 Concrete tilt-up building. *Source:* Rojahn and Reitherman (1989), ATC-20 Applied Technology Council.

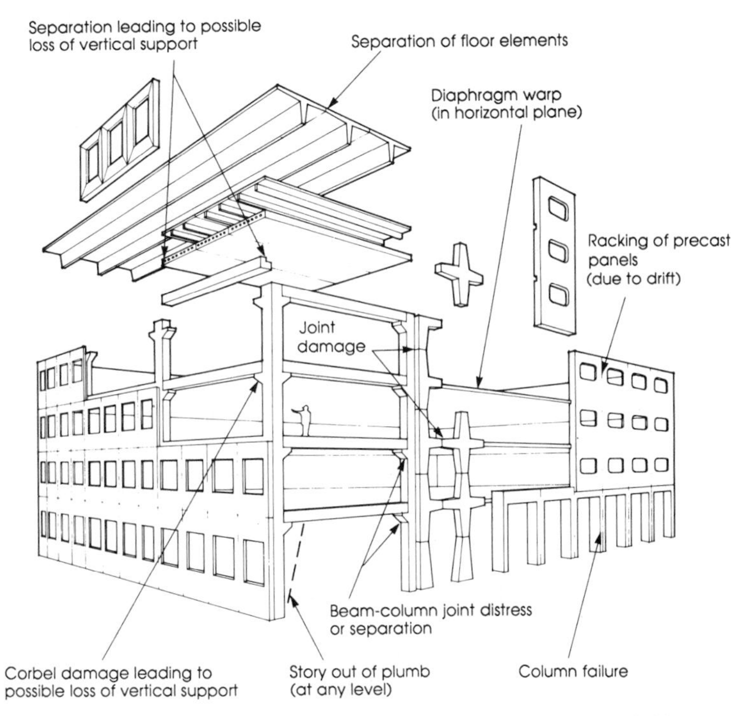

Figure 7-6 Precast concrete building. *Source:* Rojahn and Reitherman (1989), ATC-20, Applied Technology Council.

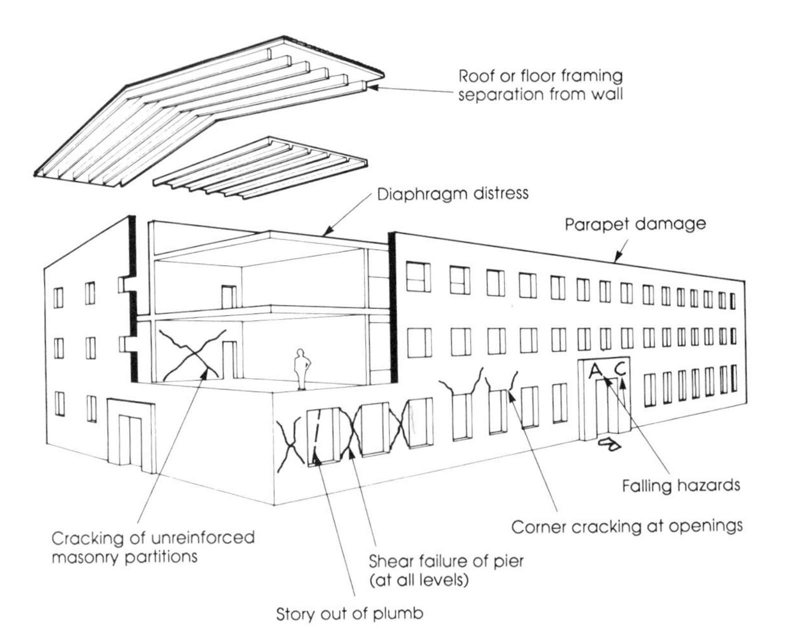

Figure 7-7 Pre-1940 reinforced concrete shear wall building. *Source:* Rojahn and Reitherman (1989) ATC-20, Applied Technology Council.

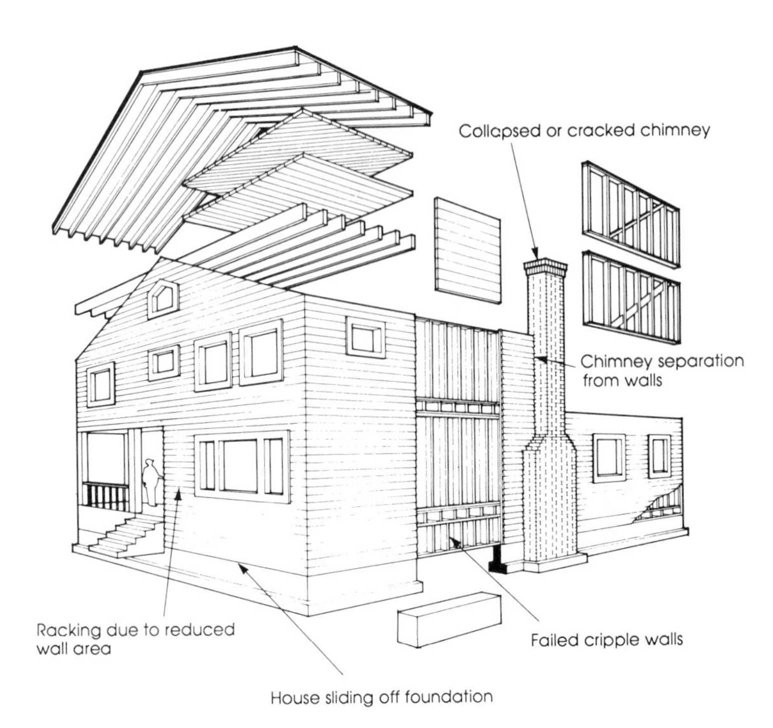

Figure 7-8 Pre-1934 light wood-frame building. *Source:* Rojahn and Reitherman (1989) ATC-20, Applied Technology Council.

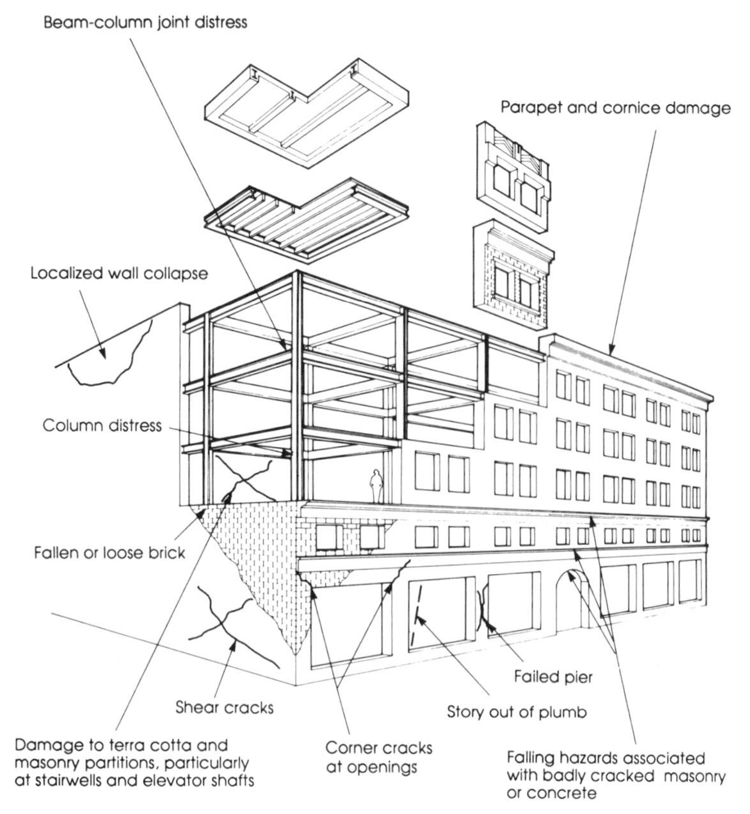

Figure 7-9 Pre-1940 steel-frame building. *Source:* Rojahn and Reitherman (1989), ATC-20, Applied Technology Council.

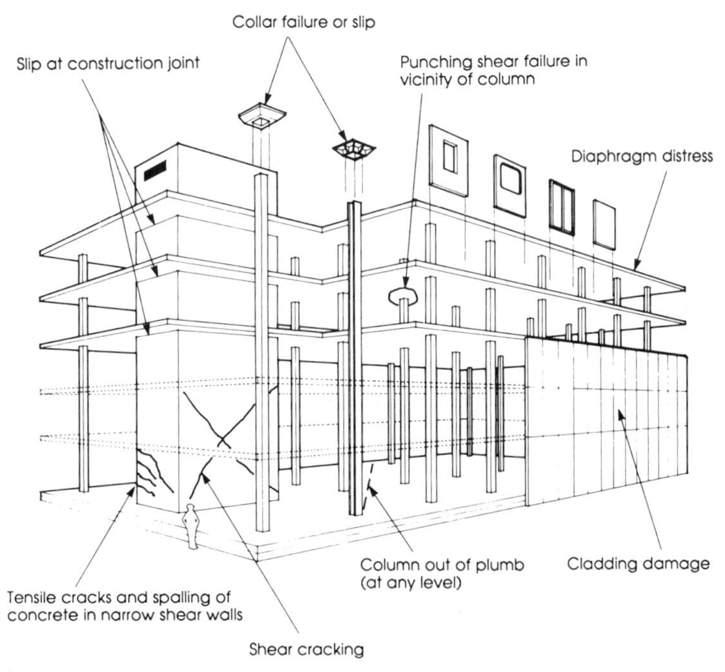

Figure 7-10 Concrete liftslab building. *Source:* Rojahn and Reitherman (1989), ATC-20, Applied Technology Council.

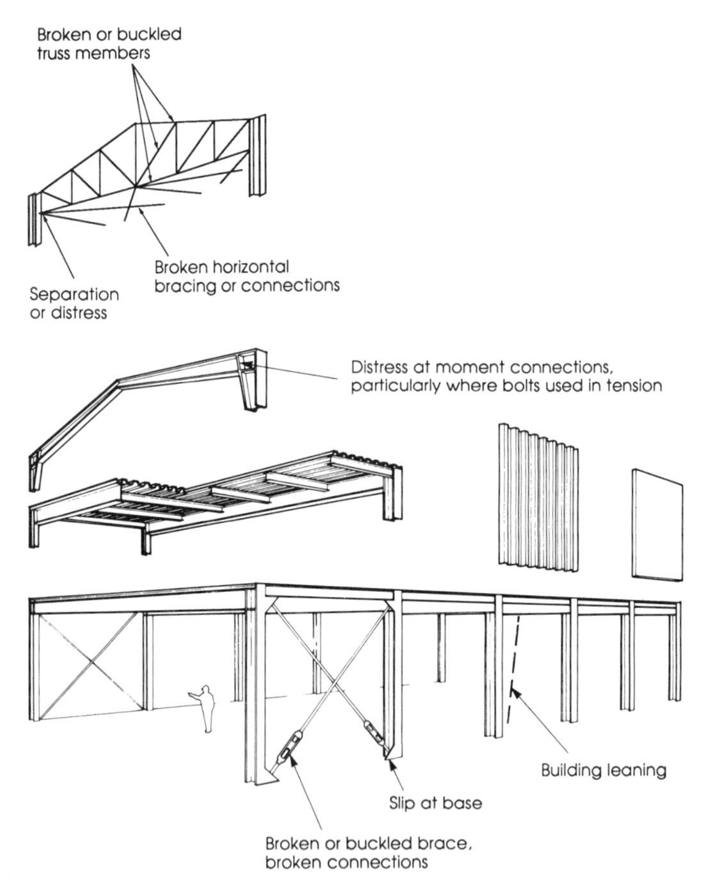

Figure 7-11 Light braced steel-frame building. *Source:* Rojahn and Reitherman (1989), ATC-20, Applied Technology Council.

Table 7-2. The buildings are shown in an "exploded" format to represent the possible use of various combinations of materials and structural components. Because of the small scale of drawings, it is impossible to render all possible arrangements or to show all of the details of a building type. The diagrams only represent typical construction characteristics and their key elements. Again, neither the descriptions in Table 7-2 nor the diagrammatic information presented should be applied to actual site-specific buildings in the evaluation of their seismic vulnerability; rather, they are to be used as representing general elements of construction.

REHABILITATION AND RETROFIT OF EXISTING BUILDINGS

Introduction

Architecturally, the seismic rehabilitation of an older building may occur for several reasons. Whether or not adaptive reuse, structural

TABLE 7-2 Potentially Vulnerable Construction Types

A. Unreinforced Masonry Bearing Walls

- A1. URM 2 stories and under
- A2. URM 3 and 4 stories
- A3. URM over 4 stories

B. Nonductile Concrete Frame

- B1. Nonductile concrete frame 3 stories and under
- B2. Nonductile concrete frame over 3 stories

C. Precast, Tilt-up and Reinforced Masonry

- C1. Liftslab construction
- C2. Tilt-up construction, pre-1973
- C3. Tilt-up construction, post-1973
- C4. Reinforced masonry
- C5. Precast concrete
- C6. Prestressed concrete

D. Pre-1940 Reinforced Concrete Systems

- D1. Under 4 stories
- D2. Over 4 stories

E. Wood Construction

- E1. Wood stud bearing wall, pre-1940
- E2. Post and beam construction, Pre-1940

F. Mixed Construction (Wood, Masonry, Concrete, Steel)

- F1. Under 4 stories
- F2. Over 4 stories

G. Steel-Frame Systems

- G1. Steel frame with masonry infill, pre-1940
- G2. Steel frame with concrete cover, pre-1940
- G3. Steel braced frame (early systems)

Source: H. J. Lagorio et al. (1986).

strengthening, historic preservation, or seismic upgrading are involved in a building's renovation, basic relationships between structural requirements and the consideration of architectural quality will always be present to influence the selection of strategies to be employed. For example, there should be careful study when the proposed solution requires structural modification to the exterior appearance of a building. In such cases, as a basic premise, seismic strengthening must consider preserving the exterior appearance of the building, since by completely ignoring it, many original architectural qualities may be destroyed.

Another major issue to be considered—in addition to life safety and other societal concerns, which are the primary motivating factors in the rehabilitation of an existing building—focuses on the economic benefits, or cost basis, of each retrofit case. Essentially, a client has six options available before making a final decision:

Option 1. Do nothing, absorb risk/liability, and maintain the status quo.
Option 2. Retrofit and strengthen existing building.
Option 3. Adaptive reuse of existing building combined with seismic retrofit.
Option 4. Remove existing building, turning property into an open space or parking lot.
Option 5. Sell or trade property and existing building.
Option 6. Demolish existing building and construct a new one.

With today's growing awareness of natural hazards by the public, the first option is clearly the least desirable, since the cost in terms of potential liabilities incurred is unacceptable, even for those with the deepest pockets. In addition, in California some local governments have legislated seismic-oriented ordinances that eliminate Option 1 by requiring the abatement of older, existing hazardous buildings, as discussed later in this section. Also, of the many strategies available for the rehabilitation of a building, adaptive reuse combined with seismic retrofit has become one of the most popular for economic reasons. In this approach, the occupancy is changed to a new use having a potential for higher income returns, such as changing an industrial warehouse into an office building or restaurant, for example.

Tax/Economic Incentives

Tax incentives to encourage rehabilitation and upgrading of older, existing buildings are based, in part, on the desirability of preventing urban centers from becoming congested areas of densely packed contemporary high-rise buildings with little attachment to the city's historic past. Other benefits include a visual enrichment of the urban fabric that constitutes many metropolitan centers throughout the country. The primary benefit, however, is aimed at the preservation and conservation of historic monuments. The advantages of seismic rehabilitation, accordingly, also fit into this rubric as other building improvements are made.

Another economic incentive for the rehabilitation of older buildings was created by federal tax legislation enacted in 1981, in which building owners and developers are allowed a tax incentive for restoring and retrofitting existing structures of identified historical value. Simply explained, the original investment tax credit typi-

TABLE 7-3 Tax Credits Allowed by Age of Building

Age of Building	Tax Credit
On *National Register*	25%
40 years (or more)	20%
30 years (or more)	15%

cally offered a one-time 25 percent write-off on income taxes in exchange for a written agreement to keep the exterior facades of older, existing buildings in their original forms.

In 1983, estimates of tax credits received for the rehabilitation of historic buildings and commercial properties, including housing facilities, reached about $600 million. Tax credits based on the age of a building (as indicated in Table 7-3) were established by the 1981 tax legislation in three increments.

This tax legislation was modified in 1986 as part of the general tax simplification package, which reduced the minimum one-time tax credit to a 13 to 20 percent tax write-off. In addition, when using this credit, accelerated depreciations cannot be claimed. All other conditions for qualifying for the tax credit remain essentially the same as those defined in 1983. However, before applying for any tax credits it is important for the architect to be familiar with the new tax credit rules and the detailed changes made under the 1986 law. (See Figure 7-12.)

TYPICAL SEISMIC DEFICIENCIES IN EXISTING BUILDINGS

Older, existing hazardous buildings may have many deficiencies that must be identified and assessed by the architect before development of final retrofit designs. Each deficiency must be evaluated carefully, since a particularly complex set of problems may lead to the economic decision that new construction may be the only viable choice. A general listing of some of the representative seismic deficiencies found in older buildings, which can be solved by architectural and engineering design strategies, include the following:

Basic configuration	Construction materials
Site/foundation	Vertical continuity
Eccentricities	Relative stiffness
Load-bearing capacity	Nonstructural elements
Connections/joints	Horizontal diaphragms
Anchors/ties/collectors	Wall openings

Department of the Treasury
Internal Revenue Service

Publication 572
(Rev. Dec. 87)

General Business Credit

Business Tax Credits

- Investment Credit
- Jobs Credit
- Research Credit
- Low-Income Housing Credit

For use in preparing
1987 Returns

Highlights

Rehabilitation credit. For property placed in service after 1986, the credit percentages are reduced to 20% for certified historic structures and 10% for buildings placed in service before 1936. The alternate test previously used to meet the external wall retention requirement is now the only test that must be met. However, the external wall retention requirement does not apply to certified historic structures.

Low-income housing credit. For tax years ending after 1986, the general business credit includes the new low-income housing credit.

Passive activities. For tax years beginning after 1986, any credits from passive activities are limited to the amount of tax generated by the passive activity. Any credits not allowed in the current tax year may be carried over to subsequent years to be used to offset tax generated by passive activities in those years. See Publication 925, *Passive Activity and At-Risk Rules,* for more information.

Introduction

This publication discusses the credits generally used by small businesses. The jobs credit and research credit that formerly were explained in Publication 906, *Jobs and Research Credits,* are now included in this publication.

Tax credits that are not covered, and references where you can find more information, are:

- Foreign tax credit—Form 1116
- Credit for clinical testing of drugs—Form 6765
- Alcohol fuel credit—Form 6478
- ESOP credit—Form 8007
- Nonconventional source fuel credit—Section 29 of the Internal Revenue Code

This publication provides information on the most common tax situations. It explains the tax law in plain language so that it will be easier to understand. However, the information provided does not cover every situation and is not intended to replace the law or change its meaning.

General Business Credit

Your general business credit for 1987 is a combination of your investment credit, jobs credit, alcohol fuel credit, research credit, and low-income housing credit plus any carryovers from other years.

There is a limit, based on your income tax, on how much general business credit you can take in any one tax year. If the general business credit you are entitled to for the year is greater than this limit, you can carry the excess to another tax year and subtract it from your income tax for that year. See *Carrybacks and Carryovers,* later.

The alcohol fuel credit is not discussed in this publication. This credit is allowed for the sale or use of straight alcohol and alcohol mixtures that are used as a fuel. See Form 6478, *Credit for Alcohol Used as Fuel,* for more information.

Limit

Before figuring the limit on how much general business credit you can take, you must figure the tax on which the limit is based. To figure your income tax for this purpose, exclude the following taxes from your income tax.

1) Alternative minimum tax.
2) Environmental tax.
3) Tax on excess benefits from a pension plan to owner-employees.
4) Tax on early distributions under annuity contracts.
5) Tax on early distributions from qualified employer or government plans.
6) Additional tax on early distributions from individual retirement accounts.
7) Accumulated earnings tax.
8) Personal holding company tax.
9) Additional tax from recoveries of foreign expropriation losses.
10) Tax on certain built-in gains of S corporations.
11) Tax on excess net passive investment income of certain S corporations.
12) Tax from recapture of investment credit.
13) Tax on nonqualified withdrawals from capital construction funds.
14) Tax nonresident aliens and foreign corporations on U.S. source income not connected with U.S. business.

Subtract the following credits from your tax to figure income tax for this limit.

1) Credit for household and dependent care expenses.
2) Credit for the elderly and the disabled.
3) Residential energy credit carryover.
4) Foreign tax credit.
5) Orphan drug credit.
6) Nonconventional source fuel credit.
7) Possessions corporation tax credit.
8) Credit for interest on certain home mortgages.

Tax liability limitations. Your general business credit is limited to the smaller of your income tax, adjusted as explained above, $25,000 plus 75% of your tax that is over $25,000, or the excess of your income tax as adjusted over your tentative minimum tax. See Publication 909, *Alternative Minimum Tax for Individuals,* or Publication 542, *Tax Information on Corporations,* for an explanation of tentative minimum tax.

For the investment tax credit limits on a C corporation, see *How to take the Credit* under *Investment Credit,* later.

Example. Your general business credit for the tax year is $40,000. Your income tax, as adjusted, is $42,500. Therefore, the amount of general business credit you can take for the tax year is limited to $38,125 ($25,000 plus $13,125 (75% of $17,500)), if you have no minimum tax.

Married persons filing separate returns figure their limits separately. Each spouse's credit is limited to the smaller of each spouse's separate income tax, as adjusted, or $12,500 plus 75% of the tax over $12,500. However, if one spouse has no credit for the tax year and no carryovers or carrybacks of any credit to that year, the other spouse can use the full $25,000 instead of $12,500 in figuring the limit based on the separate tax.

A controlled group of corporations must divide the $25,000 amount among its members.

How To Take the Credit

The general business credit for 1987 consists of the investment, jobs, alcohol fuel, research, and low-income housing credits. If you have two or more of these credits, or a carryover from an earlier year of any general business credit, you must file Form 3800, *General Business Credit,* with your tax return.

You must complete and attach the appropriate form for each credit you have and summarize them on Form 3800. You figure your tax liability limitations on Form 3800. However, if you have only one of the credits and no carrybacks or carryovers, you should file the form relating to the credit and not Form 3800.

Figure 7-12 IRS tax form entitled "Instructions for Building Rehabilitation Credit for Property Placed in Service After 1986."

Preliminary, Tentative Guidelines, 1978

One of the first documents to address the problem and establish tentative provisions for the repair and strengthening of existing buildings was the publication prepared by the Applied Technology Council (ATC) under a project funded in 1978 by the National Science Foundation (NSF) and the National Bureau of Standards (NBS), "Tentative Provisions for the Development of Seismic Regulations for Buildings." (See Sharpe and Culver, 1979.) The document also indicated that the decision to repair, strengthen, or demolish a building is based on a variety of economic considerations. Rehabilitation costs may be justified by economic benefits, such as increases in market value, anticipated lifetime, projected revenue, and/or possible tax or depreciation benefits.

Basic rehabilitation concepts identified in the publication, which are still appropriate today, include the following general recommendations:

1. Replacing or restoring damaged materials and/or faulty components of structures,
2. Increasing the thickness or size of, adding reinforcement to, and/or increasing the strength of connections and joints of individual structural components,
3. Providing additional shear walls or vertical bracing to increase capacity of lateral resistance,
4. Removing upper stories to reduce mass of the building, and
5. Shortening the period of the modified structure and increasing its response characteristics.

ORDINANCES FOR EXISTING BUILDINGS: LEGISLATION AND LEVELS OF REHABILITATION

Legislation

In consideration of liability and life safety concerns, a trend has been established on the West Coast to consider legislation to enact ordinances for the abatement of older, existing hazardous buildings with exposures judged critical to major seismic events. This is a result of recent California state legislation that requires local governments at municipal and county levels to identify, quantify, and assess older, existing hazardous buildings located within their jurisdictions. Further, they are encouraged to submit appropriate programs and plans for earthquake hazard mitigation efforts dealing with the buildings so identified. Refer to Chapter 10 for additional details on this initiative.

Whereas in the past some older structures posing earthquake hazards have been rehabilitated voluntarily by building owners, these referenced abatement ordinances mandate the seismic retrofit of older, hazardous buildings to improved levels of performance or, as an option, their demolition and removal. Some programs have now been in existence for several years.

Categories of Ordinances

Seismic ordinances are generally grouped into two categories: retroactive and triggered. Retroactive ordinances tend to be enacted by local jurisdiction at the city and county level of government. Under this category, the local government agency identifies hazards that need to be mitigated. In buildings, the options given are to strengthen, change use to a lesser risk, or to demolish.

In general, retroactive abatement codes consist of four parts. First a target is identified, which may be a class of building (e.g., all reinforced masonry buildings constructed before 1934), or an element of a building (e.g., parapets). Second, priorities for hazard reduction are determined, generally based on the level of risk associated with building occupancy types and uses. Third, strengthening requirements are established, designating appropriate levels of force resistance of the building and/or its components based on the building's existing earthquake-resistance capacity, configuration, height, and so on. And last, a time frame is developed as a scheduled period within which compliance with the ordinance must be met. The time and existing resistance capacity requirements vary according to the different risk groups assigned to building classes and types. Table 7-4 illustrates examples of building types and resistance requirements for respective risk groups.

While most retroactive ordinances require strengthening, some municipalities take another tack: identifying and documenting hazardous buildings as part of the public record. In this approach, also

TABLE 7-4 Representative Examples of Building Rating Classifications and Time Requirements for Compliance

Rating Classification	Occupant Load	Time Extension If Wall Anchors Are Installed	Time Periods for Compliance
I*	Any	1 year	0
II	100 or more	3 years	90 days
III	100 or more	5 years	1 year
	51 to 99	6 years	2 years
	20 to 50	6 years	3 years
IV**	Less than 20	7 years	4 years

Source: Scott (1985). Reprinted with permission.

Notes: * = highest priority
** = lowest priority.

known as the "embarrassment" ordinance, the building owners are encouraged to upgrade under pressure from community concerns, public exposure, and liability threat.

In a triggered ordinance, seismic requirements are instigated by government agencies prior to a building's renovation as a means to gradually improve the overall seismic performance of the existing building stock located in the community. When a building owner wishes to do substantial remodeling or renovation that exceeds a certain percentage level of cost and/or floor space additions, the rehabilitation work must include seismic upgrading in order for a building permit to be obtained. The triggered ordinances are generally accepted in most cities as a part of the local building code. Again, it should be noted that a triggered ordinance only applies to buildings undergoing renovation, whereas the retroactive ordinance applies to all hazardous buildings across the boards.

Several municipalities in California have developed seismic codes for specific application to older, existing hazardous buildings. In response to legislation sponsored by the state's Seismic Safety Commission, permitting local governments to develop rehabilitation codes that are less stringent than current code standards for new buildings, variations in "levels" of resistance requirements are approved for older buildings, rather than full code compliance being mandated. This concept has been generally accepted for economic reasons as long as the "level" selected achieves an improvement to life safety conditions, rather than settling for continuation of the original unacceptable hazard level in effect before renovation.

Code Levels for Building Rehabilitation

After it has been decided that seismic rehabilitation of an older building is in order, the next question is, What code level is appropriate for seismic force requirements to be applied for retrofit design purposes? In such cases, is full compliance with the current building code required, or are alternative levels of code compliance available?

At present, seismic resistance of an older building is still under debate, so the answer to the question is still somewhat open for assessment. Yet, for economic purposes, if no resistance value is assigned to the existing structure, retrofit costs could be substantially higher. In areas experiencing affordable housing shortages, the economic implications of this could lead to higher rental and leasing costs, which the market could not afford.

Since there is a large variation in a building's age, quality of construction, size, and structural details, a difference in seismic resistance requirements is extremely important. Most ordinances assign minimal strength values for older structures unless on-site testing data indicating higher values are available. Some ordinances require

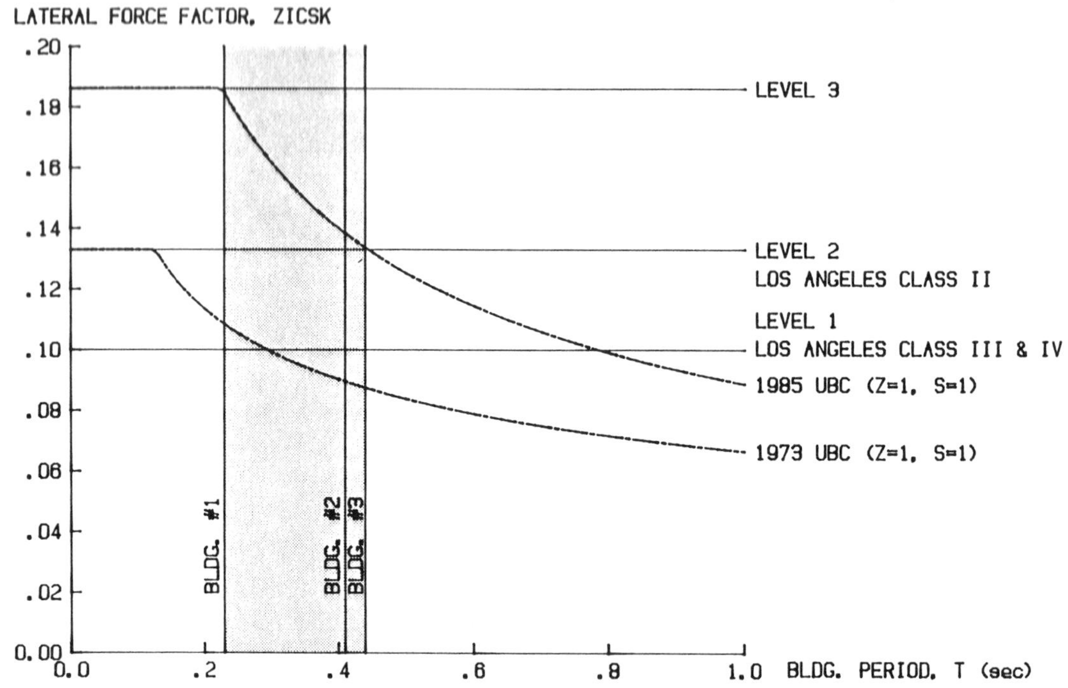

Figure 7-13 Comparison of code levels for existing URM building rehabilitation. *Source:* Wong (1987).

that all values assigned to existing materials be substantiated by testing or by establishing an adequate average working stress value. Figure 7-13 presents a comparison of code level requirements typically used for rehabilitation purposes, as found in the model Los Angeles Hazardous Building Abatement Ordinance passed in 1981. The gray zipitone area in the graph plots three case study examples of representative unreinforced masonry bearing wall buildings as an indication of their relationship to the various code/ordinance levels shown. Based on this study, it was clear that seismic retrofit costs for low-cost housing would be prohibitive if full code compliance were required in all cases.

PROTOTYPE STRENGTHENING CASE STUDIES

To decrease seismic risk in an existing building, several options are available: (1) to reduce earthquake-induced forces, (2) to change use of the building by reducing occupancy level, and (3) to make the existing building more resistant. Furthermore, these options may be applied simultaneously to any structure. The risk reduction achieved through a change in the building's use is clear: if the exposure of human activity (occupancy) in the building to earthquake hazards is lessened, so is the seismic risk of a building. For example, in 1975, to reduce the hazard to occupants, an educational institution changed a large multipurpose recreation room with an occupancy load of 50 persons used daily for about eight hours into a space for the storage of bicycles. Obviously, the threat of life safety was dramatically reduced.

Use of Base Isolation

The reduction of earthquake-induced forces can be achieved through the use of ground motion input control devices such as base isolation pads. Base isolation, which started to be effectively utilized about 10 years ago, is becoming increasingly viable as a method used in the seismic rehabilitation of older buildings. In fact, in certain cases dealing with complex structures, recent studies indicate that it can be more cost-effective for strengthening and retrofitting a building than some of the traditional approaches. There are many limitations to the effective use of base isolation devices; for example, they cannot be used on high-rise structures because building excitation problems may produce excessive deformations. However, even though base isolation cannot be used for all building classes and construction types across the boards, it remains a viable system worthy of further consideration.

Representative Examples: Strengthening Methods and Techniques

In this section, only the option of strengthening through the addition of rigid elements will be discussed in detail, since this approach is currently the most utilized and does not involve the more experimental techniques such as base isolation and other special energy absorption systems. Strengthening methods using the addition of rigid shear-resisting systems must give consideration to the use of (1) vertical shear wall elements, (2) horizontal shear-resisting diaphragms, (3) architectural configuration and characteristics, (4) foundation and site conditions, (5) overall structural integrity, and (6) coordination with architectural considerations. It must be noted that these options may not be applicable to all potentially hazardous structures, as many buildings have unique characteristics requiring individual analysis rather than prototype solutions.

VERTICAL SHEAR-RESISTING ELEMENTS

The various types of vertical shear/bracing elements commonly used may be placed into three categories: (1) shear walls, (2) stiff frames, and (3) diagonal bracing. Each type in turn can be constructed out of different materials. Vertical shear elements have a great impact on architectural planning, because they can affect space planning flexibility in terms of circulation patterns, plan lay-

Mechanical advantage of configuration in "X" and "K" braces

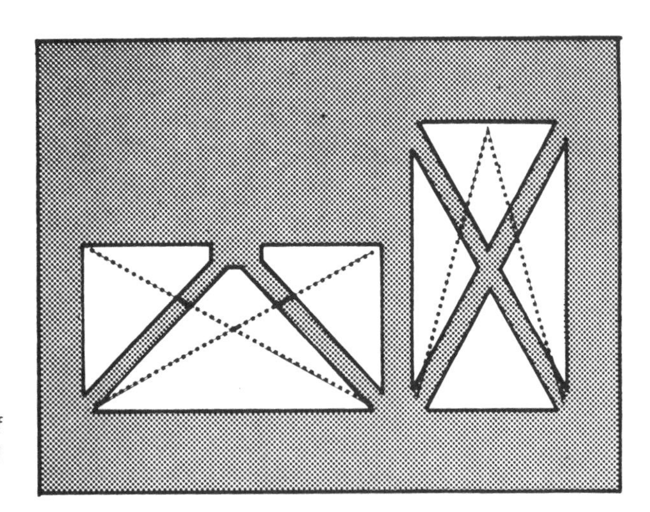

Door penetration possibilities of "K" braces

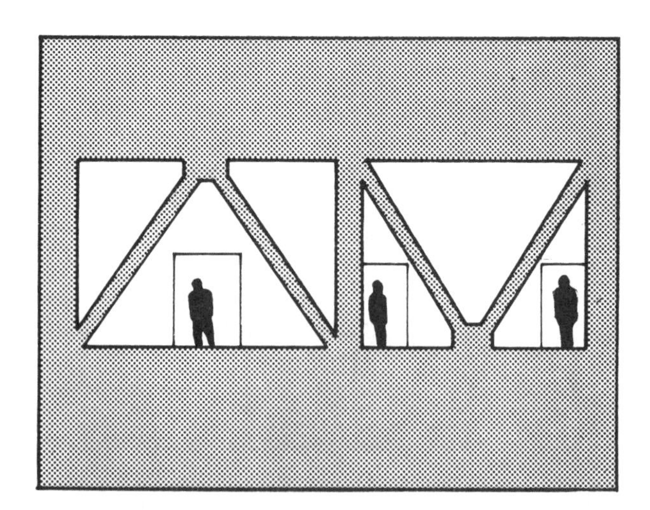

Existing floor can be cut to allow for pour-in-place concrete shear wall.

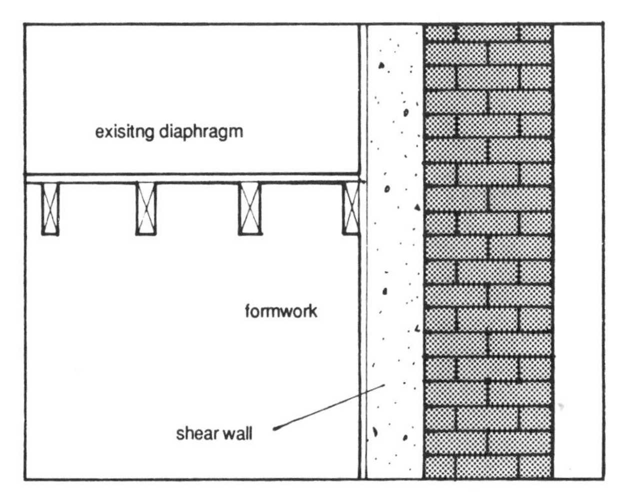

Figure 7-14 Examples of vertical bracing systems. *Source:* Wong (1987).

outs, or wall openings, for example. Second, once a vertical shear element is selected, other components are usually selected to be compatible with the vertical system. For example, when a steel braced frame is used as a vertical bracing system, steel chord members will probably be employed for diaphragm edge strengthening because of the economical advantage of using one type of material throughout a renovation project.

Figure 7-14 shows three examples of typical vertical shear/bracing methods commonly used in the seismic retrofit of older hazardous buildings.

HORIZONTAL SHEAR-RESISTING ELEMENTS

Horizontal shear-resisting elements, also known as diaphragms, refer to stiff horizontal elements that transfer loads from the walls or floors of a structure to the vertical shear-resisting elements, which in turn distribute them to the foundations. Diaphragms and collectors are two main components of commonly used horizontal shear elements. The primary structural role of the diaphragm is to keep the structure free from excessive distortion and/or torsion.

Although serving an important structural need, use of horizontal diaphragms and collectors for strengthening purposes is often of little architectural planning concern, except for decorative ceilings and floors. The modification of existing floors, ceilings, and roofs into elements stiff enough to transmit seismic loads to vertical shear-resisting elements usually does not create special architectural problems, except when dealing with the preservation of architectural finishes, or where there are many vertical penetrations which would affect the integrity of the diaphragm.

Figure 7-15 indicates typical locations for the installation of horizontal diaphragms at roof, ceiling, and floor levels.

CHANGES IN CONFIGURATION CHARACTERISTICS

In certain cases the very configuration of a building, as Arnold and Reitherman indicated in their book, *Building Configuration and Seismic Design* (1982) may represent a deficiency in an existing building's capacity to resist seismic forces. Strategically placed, changes to the vertical (elevation) or horizontal (plan) configuration of a building can improve its seismic performance either by: (1) stiffening the structural system's capacity; and/or (2) simplifying the building's configuration by separating it into simple, separate masses through the use of seismic separations or joints.

Figure 7-16 presents possible prototype solutions showing a horizontal (plan) change made in a building's configuration and vertical (elevation) changes made to a building's vertical continuity to improve its general seismic performance.

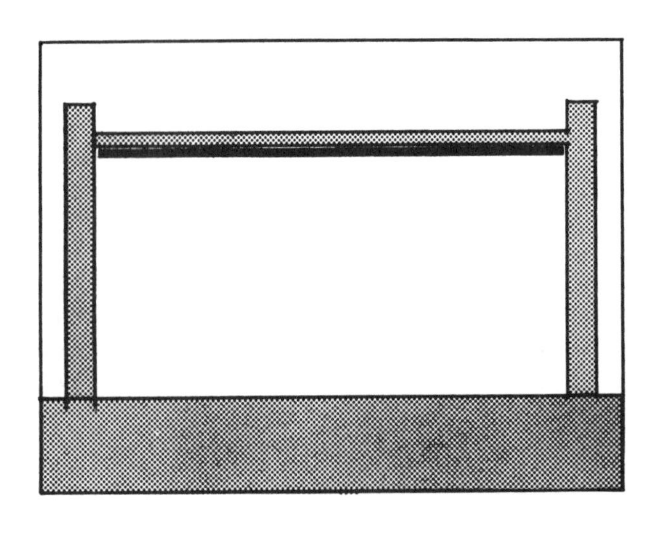

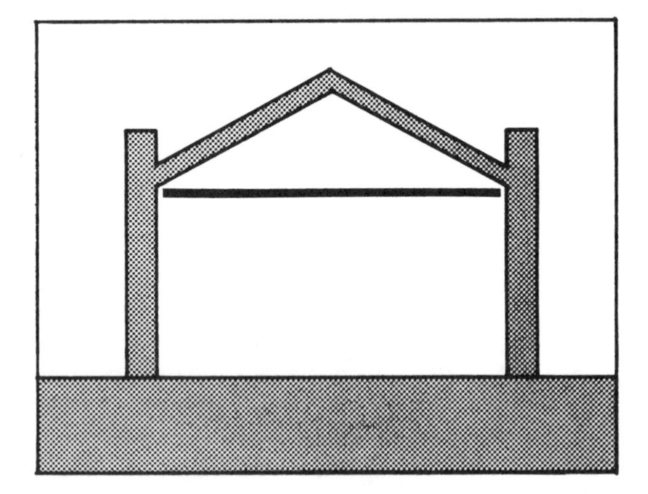

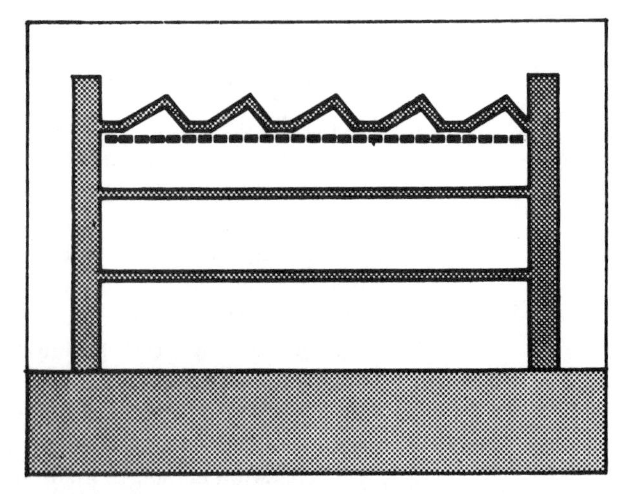

Figure 7-15 Examples of horizontal diaphragms at (from top to bottom) roof, ceiling, and floor levels. *Source:* Wong (1987).

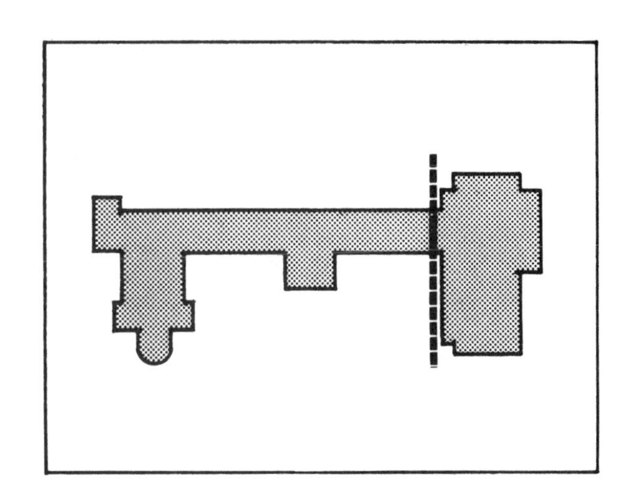

Seismic separation in the retrofit of a school building

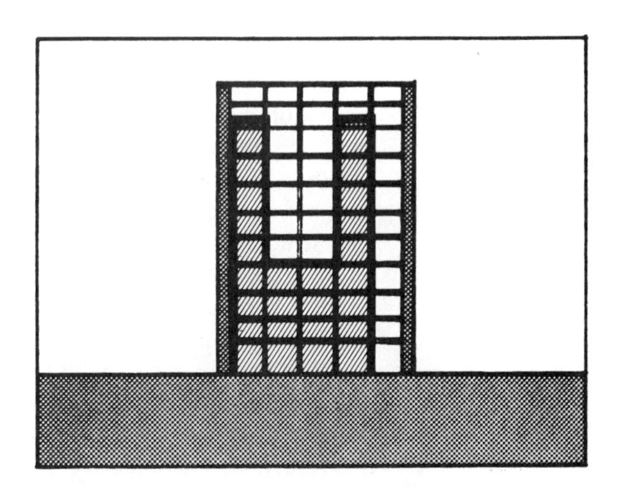

Fork configuration of shear wall in San Francisco building

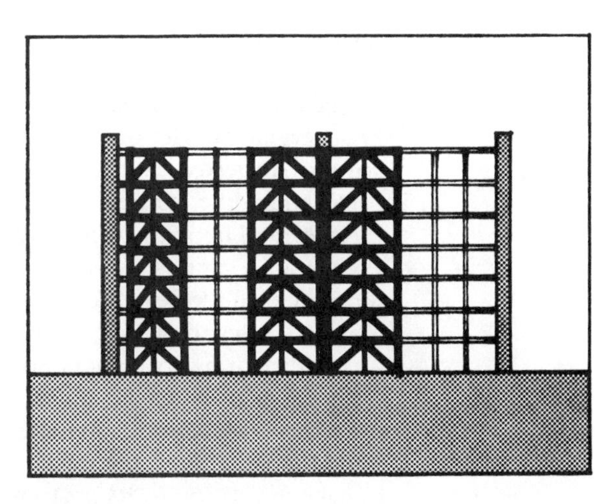

Example of tower configuration of the shear elements

Figure 7-16 Examples of Solutions to a building configuration problems. *Source:* Wong (1987).

RELATIVE RIGIDITY OF ADJACENT BUILDINGS IN METROPOLITAN CENTERS

Individual Building Context

The strengthening and seismic retrofit of an individual building must be approached with great discretion, since each building presents a different set of problems. Additionally, each building will have its own behavior under earthquake force excitation. Unique problems will typically appear for each building under consideration owing to the great number of variables we are dealing with.

The architect will quickly find that there are no simple solutions: each building will have to be addressed on an individual basis from the very start. However, considerations for a single isolated building are very different from those involved for a building in a neighborhood setting of an entire block, where multiple buildings sit side by side; here, the variables increase significantly. In Chapter 8, which presents the total urban context of earthquake hazards mitigation more thoroughly, the implications of this for the entire existing building stock will be explored in greater detail.

When dealing with an older structure on a block full of buildings, the characteristics and anticipated performance of all the adjacent structures must be assessed. As seen in the 1985 Mexico earthquake and the October 1989 Loma Prieta earthquake, corner buildings are particularly vulnerable to damage, as they have nothing to lean on at the open street side (see Chapter 12).

When designing a new building to be located next to an existing building, the problem may be solved by providing a seismic separation, as will be explained in Chapter 8. But in terms of seismic rehabilitation, an objective solution is difficult when a group of older buildings of diverse rigidity have been built side by side and to the property line.

Unfortunately, there is no simple solution, and the matter is still being debated by design professionals. Some have asked the question: "What value is it to upgrade an individual URM building when in fact all the buildings on the block not retrofitted might act in concert and impact the strengthened building negatively?" One solution suggested is that it might be more appropriate only to strengthen the corner buildings in every block, almost as buttresses, and hope that they will support the rest from collapse. Another suggestion is that if possible, all seismically hazardous buildings in each block be tied together to act as a unit laterally.

It is unclear how a city could regulate such actions in a society of individual property owners. Still, it is important to note that there is a great deal of debate on the part of design professionals as to whether the seismic upgrading and strengthening of only one building on a block might have an overall adverse effect in an earthquake in relation to its unstrengthened neighbors, or vice versa.

TABLE 7-5 Comparative Costs of Required Levels of Seismic Strengthening (For Cost/Sq Ft in 1986 U.S. Dollars)

Building	Wall Anchorage Only	Level 1	Level 2
No. 1 (2 stories)	$6	$13	$21
No. 2 (4 stories)	$6	$16	$23
No. 3 (5 stories)	$6	$21	$25

Source: Wong, CEDR, (1987), and Comerio, et al. (1987).

COST RELATIONSHIPS

An NSF-funded research project completed in 1987 by Dr. Kit Wong for the Center for Environmental Design Research (CEDR), University of California at Berkeley, reveals that variations in retrofit design solutions that most dramatically impact costs are primarily those related to: (1) improvement in the performance of a building's overall characteristics (e.g., height, configuration, rigidity, etc.); and (2) the level of strengthening mandated by code/ordinance requirements. Refer to Table 7-5 for increases in retrofit costs associated with building height and levels of strengthening for three URM building case study examples.

Additional cost figures for strengthening URM buildings are also covered in another publication issued by the Federal Emergency Management Agency (FEMA) in Washington, D.C., under a contract awarded to Englewood & Hart, Consulting Engineers, Los Angeles, California.

Figures 7-17 through 7-19 indicate three typical URM rehabilitation details commonly suggested for use in seismic upgrading as a means of tying the building together so that it will act as a unit: (1) the installation of joist anchors, (2) out-of-plane wall bracing, and (3) parapet bracing/strengthening.

SUMMARY

Finally, in dealing with the seismic rehabilitation of older buildings on a cost-effective basis, it appears that the greatest payoff, or maximum benefit, per dollar spent on effective earthquake hazards reduction options available to the design professional are represented by seven approaches:

1. Tie/anchor building together to act as a unit or total system,
2. Brace building (vertical shear walls or frames) to improve

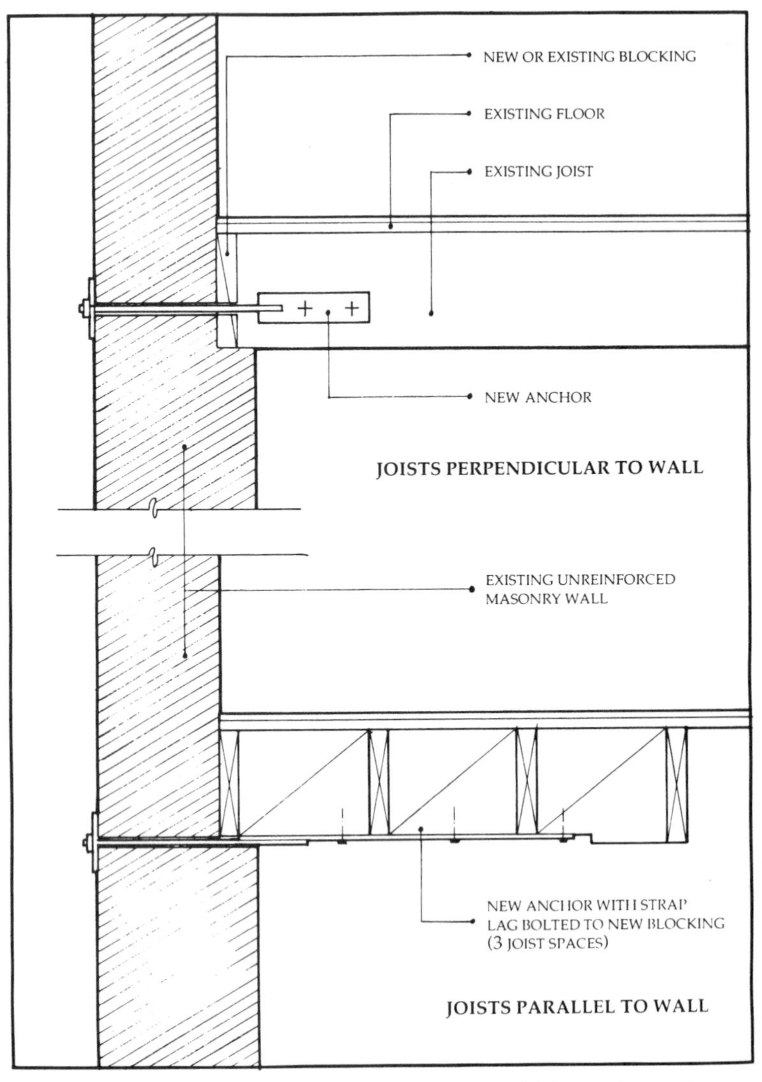

Figure 7-17 Example of solution to installation of joist anchors. *Source:* Wong (1987).

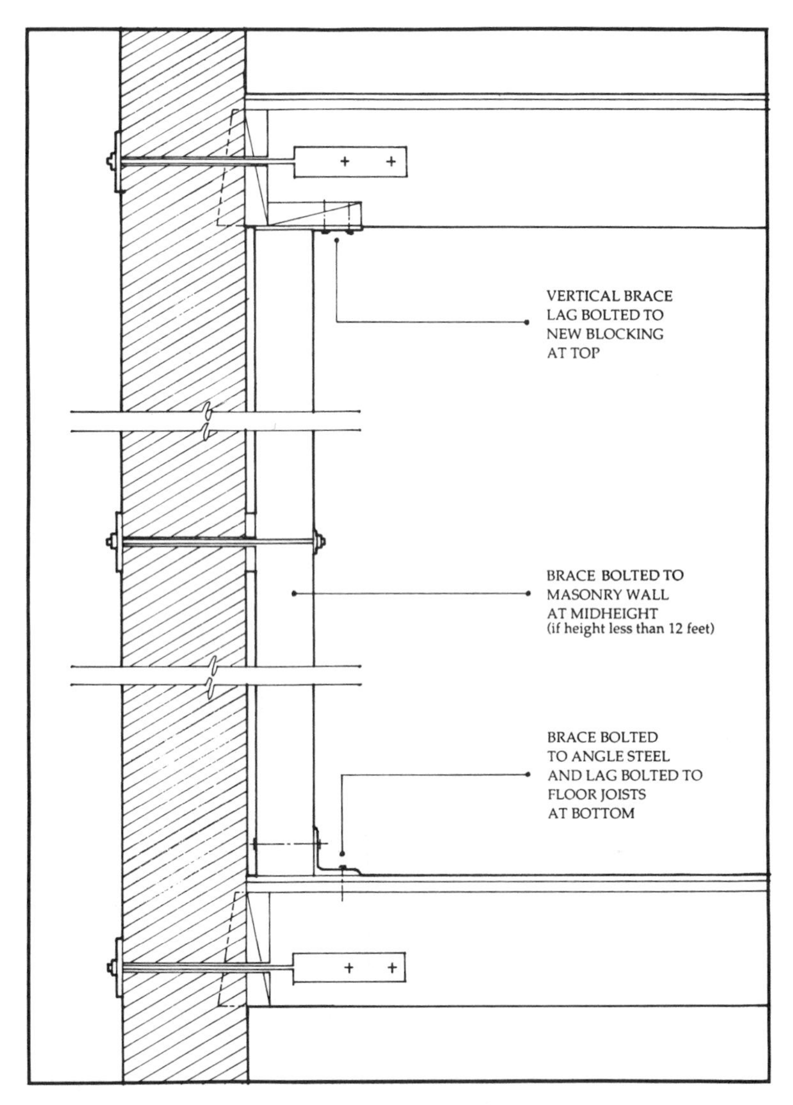

Figure 7-18 Example of solution to out-of-plane wall bracing. *Source:* Wong (1987).

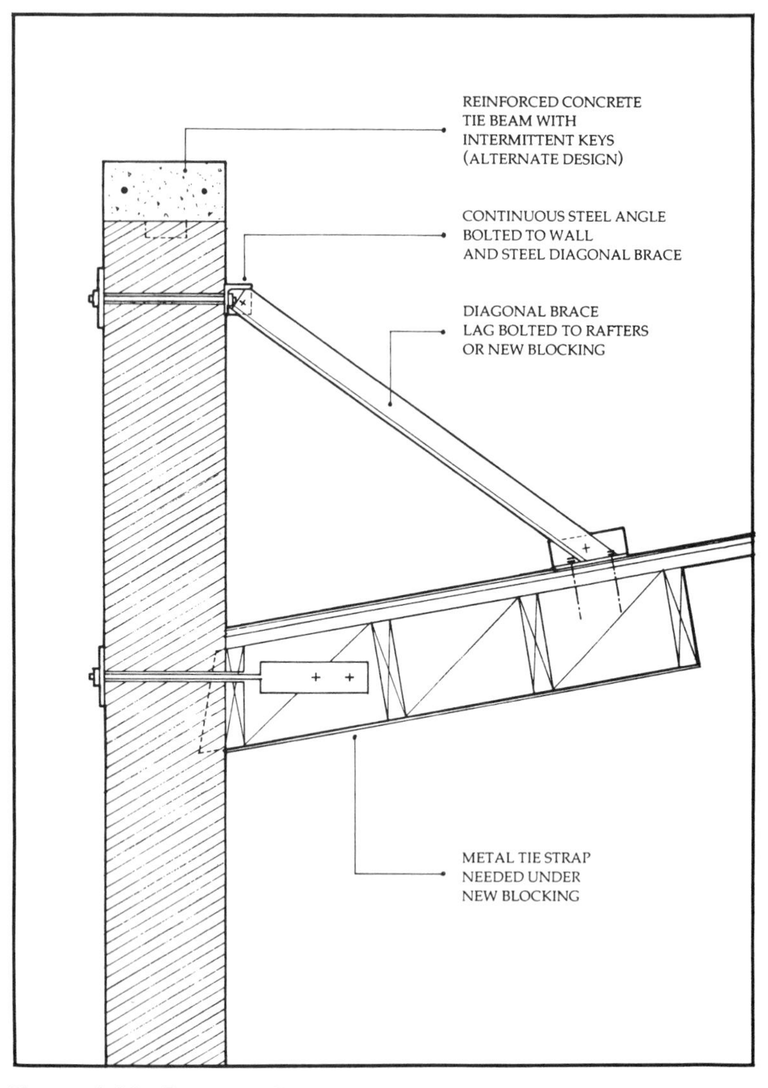

Figure 7-19 Example of solution to parapet bracing. *Source:* Wong (1987).

Figure 7-20 Cutaway view of a typical cylindrical base isolation pad element after testing *Source:* Earthquake Engineering Research Center (EERC), University of California at Berkeley.

stiffness and rigidity with emphasis on lower floors/soft stories,

3. Extend bracing elements to upper floors,
4. Provide horizontal diaphragms,
5. Decrease building mass and/or loads,
6. Improve site conditions, and
7. Decouple building from ground motions, by using base isolation. (See Figure 7-20 for an illustration of a cutaway view of a typical base isolation pad element.)

CHAPTER

8 URBAN PLANNING AND DESIGN

Although urban planning and design has often been cited as having a potential role in earthquake hazards mitigation activities, in the United States it has had a relatively short history in reducing seismic risk. There is, however, evidence that recent achievements in various seismic safety programs are beginning to motivate planners to consider earthquake hazards reduction efforts in their work when dealing with projects located in areas of high seismic risk. Interest in urban planning issues surfaced dramatically after seven counties in the greater San Francisco Bay metropolitan district were declared federal disaster areas following the October 1989 Loma Prieta earthquake (see Chapter 12 for a detailed accounting of this seismic event).

In the United States as well as in other highly developed countries, which were early beneficiaries of the industrial revolution, large-scale urbanization began in the latter half of the nineteenth century. Since that time, there has been a steady migration of people from rural areas to cities and metropolitan hubs. In the past few decades the number of major cities is still increasing, with most of the population growth taking place in existing, well-established urban centers. It has been predicted that by the year 2000, more than half of the world's people will be living in cities with a population of 250,000 or more, and that there will be about 408 cities having more than a million inhabitants. Urban planning in earthquake-prone areas should recognize the dynamics of growth and change that are taking place in order to develop an environment less vulnerable to future earthquakes.

High-density urban developments around the world exhibit great diversity. Each has developed historically under different circumstances, in different geophysical settings with diverse climates, with different architectural styles and construction methods based on different socioeconomic motivations, while attracting different inhabitants with various backgrounds. All these urban centers differ in age, the oldest having existed for centuries, the newest having been developed in recent decades. In the United States, they range from large incorporated cities with high concentrations of populated centers covering square miles in area, such as Boston, Seattle, St. Louis, and Salt Lake City, to metropolitan giants, like New York, Chicago, and Los Angeles.

The same pattern of diversity exists in other developed countries. What all these cities have in common, however, is found in the underlayment of geophysical characteristics that compose the upper layers of the earth's crust upon which they stand. This commonality is dramatically true for those cities located in those zones of high seismic risk found in many regions of the world. Owing to the very nature of their complexity, these cities remain extremely vulnerable to damaging earthquakes. Three most important variables that have the greatest effect on the vulnerability of an urban center area are its location, population density, and existing building stock. Very simply put: the less people and/or buildings in the area, the less the threat to life safety from seismic events.

GENERAL URBAN CHARACTERISTICS

The vitality of a major city is recognized as being crucial to a region's economic strength and quality of life. Large metropolitan areas are, moreover, more than just hubs of job opportunities, communication networks, recreation/entertainment outlets, and commercial enterprise. They also represent centers of learning, cultural development, housing, social services, and resources and financial support, the combination of which creates a sensitive fabric binding everything together into a functional, operational, and delicately balanced environment. The glue that binds everything together is exemplified by the existing building stock and physical infrastructure. Yet, it is exactly this binding element that is the most vulnerable to earthquakes, and one that denotes a significant economic, physical, and social investment. In extreme cases, on strictly economic terms, extensive earthquake losses to the physical infrastructure of a major metropolitan center may impact a nation's gross national product as funds are diverted to recovery efforts.

ECONOMIC AND POLITICAL IMPACTS

Within the last 15 years, several earthquakes throughout the world have impacted major cities severely enough to affect the economic and political stability of the region. After the October 1989 Loma Prieta earthquake in the San Francisco Bay Area of California, where damage estimates ran as high as $8 billion, the governor approved a temporary increase in the state sales tax from 6.5 to 6.75 percent to help pay for the losses and cover reconstruction costs. Under such drastic circumstances, it has been necessary for developing countries to support postearthquake recovery efforts with outside resources, such as those provided by the World Bank, UNESCO, the International Monetary Fund (IMF), and other international funding units. The recovery of metropolitan centers after a severely destructive earthquake has become a global problem tied to an international scale of economics, assistance, and redevelopment.

Recent data indicate that large, high-density, congested cities located in regions of high seismic risk, of which there are many, may well face dependency on special taxation or outside assistance during the immediate postearthquake recovery period. Recovery and reconstruction after the 1976 Tangshan earthquake, in which the Chinese authorities refused all outside assistance, including the Red Cross, may very well be the last time a country can afford the luxury to "do it alone." The 1985 Mexico earthquake and the 1988 Armenia earthquake in the USSR (see Figure 8-1) both confirmed this trend toward seeking and accepting international assistance.

Figure 8-1 Typical precast concrete building damage in Spitak after 1988 Armenia, USSR, earthquake. *Source:* Arnold (1989).

WORLDWIDE URBANIZATION

Urban growth has been unprecedented since World War II. An interesting study by Professor Barclay G. Jones, Cornell University, (see Jones, 1989), indicates that in 1950 there were 78 cities with populations of more than a million people, with the number, as indicated earlier, now expected to increase to 408 by the year 2000. Furthermore, his study discloses that the most dramatic emergence of large metropolitan concentrations and increases in urban populations are expected in regions of the world other than those with which large urban centers were previously associated; these trends are anticipated to be more notable in Eastern countries, Southeast Asia, and Latin America. By the year 2000 it is estimated that the urban populations in developed regions will have increased to 1 billion, whereas it is expected that it will double to 2 billion in less developed areas.

With this rapid urbanization of the world, Jones states that "populations are becoming more concentrated in the sense that larger percentages of total populations are in urban centers and smaller ones in rural areas. . . . The extraordinary growth of large cities and enormous metropolitan areas as a result of tremendous urbanization of the world population means quite simply that more major cities and urban populations are subject to the devastations of disasters than ever before." He also points out that the "most propitious locations for urban centers may be in the most disaster prone sections of the region."

In 1989–1990, the largest city in the world is still Mexico City, whose origins date back to the Aztec culture and whose current population is well over 20 million. It is the capital city of a country in which urbanization is ever-increasing at an intensified rate. Using Mexico as an example, Table 8-1 is illustrative of the rate of urbanization that is most probably taking place in many other cities throughout the world. As indicated, by the year 1990 it is expected that 90 percent of the population in Mexico, representing about 89 million people, will be living in urbanized areas. This model would also be true for many other rapidly growing countries, east and west.

The data in Table 8-1 indicate an important trend relative to seismic safety because the risk exposure to earthquake hazards increases exponentially in highly populated urbanized centers, in

TABLE 8-1 Mexico: Population and Rate of Urbanization, 1930–1990

Period	Urbanization Rate	Population
1930s	20%	16.6 million
1970s	80%	62.3 million
1990s	90%	89.0 million

Figure 8-2 Destroyed railroad rolling stock fabrication plant, 1976 Tangshan, China, earthquake. *Source:* Ministry of Construction, People's Republicof China.

contrast to rural areas, where risk is generally low. And this is one of the factors complicating the problem for the planners, because earthquake source data clearly pinpoint metropolitan centers as the places in which the greatest losses, whether economic or social, will occur. Accordingly, on a worldwide basis, the potential exposure to earthquake hazards in well-developed cities will increase and continue to become acute unless comprehensive mitigation efforts are undertaken. For example, the 1985 earthquake that caused extensive damage in Mexico City resulted in 20,000 deaths and devastated approximately 5700 buildings. Of this total, 485 buildings suffered total or partial collapse, while 271 experienced serious structural damage. (See Figure 8-2.) In 1976 in the city of Tangshan, China, which had a population of 1.2 million, a major earthquake leveled all the buildings in the city and killed 250,000. The 1988 Armenia earthquake is estimated to have killed 50,000.

THE URBAN CONTEXT

Rather than highlighting the seismic performance of a single building type or the classification of individual structural systems, the urban context of seismic safety focuses on the unique characteristics of the city which makes urban hazards mitigation a distinctive problem. Because of its size and complexity, the city, as noted above, is an elaborate network and unveils a extensive set of services upon which urban populations depend for survival. In order to understand its vulnerability to seismic events, interrelationships at all levels within the urban context require examination.

Urban hazards mitigation efforts require a comprehensive policy and integrated process approach. As urban planning and design in consideration of seismic hazards reduction program depends on the overall performance of the urban infrastructure, urban hazard mitigation efforts demand an understanding of the growth and development process. It must be recognized as a dynamic evolutionary process to be approached as a shared responsibility encompassing five interrelated considerations:

1. Microzonation and land use planning,
2. Essential, critical emergency facilities and lifelines,
3. Existing building stock,
4. Social concerns of public health and safety, and
5. Long-term economic impacts.

MICROZONATION AND LAND USE PLANNING

Over time, the location, frequency, and size of an earthquake are critical elements in the potential impact that a seismic event will have on the life safety of urban populations. Consequently, one of the first steps in determining appropriate urban earthquake hazards mitigation measures is to focus on the city's physical location as an indication of its seismic exposure. This is fundamental, for obviously if a city is *not* located in a high-risk seismic zone, no problem exists. And in fact there are some places that have little or no seismicity; for example, parts of France, central west Africa, eastern Brazil, and some southern states in the United States are recognized as places of low seismicity.

On the other hand, many well-known major cities of the world are located in areas where the effects of damaging earthquakes are commonplace: Tokyo, San Francisco, Los Angeles, Anchorage, Mexico City, Naples, Algiers, Guatemala City, and Caracas, to mention a few. In locations such as these, where simple relocation to a site of lesser seismicity would be out of the question, other solutions must be sought.

Table 8-2 indicates the seismic zone location and population figures of selected U.S. cities, along with a numerical evaluation of the seismic risk they face; the higher the number, the greater the risk.

In the locations shown in Table 8-2, again, simple relocation to

TABLE 8-2 Seismic Zone Location and Population of Selected U.S. Cities

Metropolitan Area	Seismic Zone (UBC)	1986 Population
Anchorage, Alaska	4	235,000
Los Angeles, California	4	3,259,000
San Francisco, California	4	1,015,000
Memphis, Tennessee	3	653,000
Seattle, Washington	3	486,000
Salt Lake City, Utah	3	158,000
Boston, Massachusetts	2A	574,000
St. Louis, Missouri	2A	426,000
Charleston, South Carolina	2A	68,000
Kansas City, Kansas	2A	441,000

another site of lesser seismicity would be out of the question, and other solutions are needed. One of these is to adopt suitable land use practices in all areas in general, but particularly in older areas already subject to rehabilitation and/or in new areas where urban growth and expansion is occurring. It is easier to adopt new ordinances for land use purposes in areas where opportunities for change already exist and new developments are taking place.

For appropriate land use recommendations to be developed in urban areas of high seismic risk, it is first essential to evaluate the geologic hazards that exist in the region. Data are collected on all existing hazardous geologic conditions, such as poor soils areas (including those of potential liquefaction), landslides, tsunami coastal areas, and active fault zones. This information is then plotted on a map to form what is technically called a "microzonation map," or seismic zonation map, for use in land use planning.

This map may then be used for making decisions regarding guidelines and recommendations on land use planning based on the hazards indicated. Such recommendations may lead to a zoning ordinance passed by the local government to limit construction, development, or specific "land use" of an area if the use is incompatible with the hazard found there. As a general example for comparative purposes only, Table 8-3 presents a generic model of a simplified cursory matrix that was developed solely as a representative target for urban planning policy to reduce the effects of earthquake hazards. By correlating geologic hazard data plotted on a microzonation map with appropriate land use functions, it demonstrates one of many approaches that may be used in addressing the problem. Societal needs and local conditions of each urban setting, of course, would require modification of the matrix according to local government policy. Obviously, for specific urban centers in areas of lower or higher seismic risk, the matrix would be

TABLE 8-3 Simplified Urban Planning Matrix: Correlations of Appropriate Land Use Policy With Geologic Hazards

Exposure to Geologic Hazard	Risk to Life Safety	Potential Land Use
Very high	Very high	Open space, agriculture
High	High	Storage, warehouses
Moderate	Moderate/low	Single-family, 1-story, detached housing
Moderate	Low	Standard high-occupancy buildings
Low	Very low	Special high-occupancy buildings
Very low	Insignificant	Critical facilities
Insignificant	Negligible	Toxic chemical plants, large dams, emergency service facilities

more extensive and also modified to become less or more restrictive, depending on the perceptions of seismic risk levels that exist in the region.

Another approach to appropriate land use planning in seismic areas is to base decisions on the annual probability of earthquake occurrence. Table 8-4 illustrates such a model.

Examples of critical service facilities include some medical facilities, local government response units, essential lifelines, and similar structures. Large-scale potential disaster impact functions may include toxic chemical plants, dams, and similar functions whose catastrophic failure would affect heavily populated metropolitan regions.

Obviously, Tables 8-3 and 8-4 could be expanded and refined to include additional safety levels, functions, and land uses as needed for the occasion. Also, one level of exposure to geologic hazards (e.g., "high") may be matched with another level of risk to life safety (e.g., "low") to derive an appropriate land use function for a particular correlation required by community priorities. In any event, as a means of reducing earthquake hazards, the two tables are included as rough models that may be used to establish rational local government policy on land use planning guidelines.

Throughout the country, many city planning provisions and building ordinances normally do not take into account the possibility of surface faulting due to earthquake. Responsible design professionals and planners have on occasion successfully persuaded clients to consider alternative sites only to have the abandoned site developed by others who may have been uninformed of the hazard.

Currently only the state of California has a fault-line hazards act in effect that mandates local governments to take fault-line hazards into account before land use plans are realized and site develop-

TABLE 8-4 Land Use Planning Matrix and Relative Correlation With Annual Probability of Occurrence

	General Levels of Risk		
Annual Probability	For Economic Loss	For Life Loss	Relative, Potential Land Use
1:10	High	High	Open space, agriculture, nonstructural use
1:100	Moderate	Moderate	Single-story housing, recreational use
1:1000	Low	Moderate	Low-rise residential or low-density commercial use
1:10000	Low	Low	High-occupancy structures
1:100000	Very low	Very low	Critical facilities, chemical plants, large dams, etc.
1:1000000	Negligible		

Source: Office of Earthquake Studies, U.S. Geological Survey (1984).

ment begins. In large cities outside California, it is entirely conceivable that major structures will continue to be constructed in earthquake fault-zone areas. Even in California some important facilities, such as acute-care general hospitals, public schools, and portions of the San Francisco Bay Area rapid transit system (BART), prior to enactment of fault-zone hazards legislation in 1974, were located across known and active fault zones without special precautions having been taken. In all probability, the pressures of urban growth, economic expediency, and increased population needs were responsible for such land use planning decisions at the time. An explanation of the purpose, use, and objectives of fault-zone hazards legislation follows.

Special Study Zones

An example of how earthquake hazards mapping may be used for planning purposes is found in the enactment of legislation governing geologic "special study zones" for construction purposes. As used in earthquake hazards mitigation programs, special study zones encompass fault-line zoning of active known faults as a land use control mechanism. Fault-zone mapping was completed in California after passage of the Alquist–Priolo Special Study Zone Act following the 1971 San Fernando earthquake.

No major facility can be planned for or located on any site designated in the special study zone without a thorough geological investigation, including trenching of the site if necessary, to preclude building major structures directly on the fault line. By law, provisions of the act must be met before any building permits are issued for construction on any site within the special study zone.

The only exception is made for simple, single-family detached residences, or duplexes, designed for private use. In effect, special study zone mapping is one form of the microzonation process that restricts construction along earthquake fault zones to certain structures. Certainly, the location, siting, and construction of critical emergency service facilities (described in the next section) such as major acute-care hospitals must conform to the provisions of special study zone legislation enacted for land use purposes.

ESSENTIAL, CRITICAL EMERGENCY LIFELINES AND FACILITIES

Just as the basic structure of a building sustains the building system, the interdependency of essential, critical emergency lifelines and facilities sustains the functional and operational integrity of a metropolitan center. Even if we don't consider earthquakes, without electric power or a water supply cities would cease to function, since the fragility of our contemporary urban centers is that sensitive; a major earthquake would compound the problem enormously.

In terms of urban planning and design relative to earthquake hazards reduction, certain facilities should be immediately available after a damaging earthquake if a city is to survive and successfully mount postearthquake recovery efforts with dispatch. The ideal is to have these critical urban "lifelines" in a position to continue services and be fully operational following a disastrous earthquake and not become an additional liability during recovery efforts.

Among all the public service facilities essential to the viability and integrity of a community, there are several that are considered most critical to continued operations during and after a severe earthquake (see Table 8-5). When in perfect working order, they are taken for granted, but when they malfunction or are impaired by a damaging earthquake, all normal urban activities can come to a grinding halt amid great frustration and consternation. For example, the collapse of the Highway I-880 Cypress Street overpass and the deck of the Oakland–San Francisco Bay Bridge during the Octo-

TABLE 8-5 Essential, Critical Emergency Facilities and Lifelines

Major hospitals	Dams, water supply and treatment plants
Ambulance services	
Fire stations	Electric power plants
Police stations	Communication centers
Airports	Interstate transportation
Pipelines	Freeways and bridges
Blood banks	Emergency services offices

ber 1989 Loma Prieta earthquake destroyed vital transportation links in regional traffic patterns which affected the trucking industry and the lives of over a million commuters. (Chapter 12 contains a detailed accounting of the 1989 Loma Prieta earthquake.)

Structures such as police and fire stations should continue to function during and after an earthquake to fight fires that may erupt following the event and to facilitate search and rescue efforts. Several examples exist wherein severe damage to fire stations precluded the use of critically needed equipment during the immediate postearthquake recovery period. When police and fire stations do survive the earthquake and remain functional, the injured still must be taken to medical facilities for treatment and recovery. If hospitals themselves have been damaged or rendered nonfunctional, and are unable to receive and care for the injured, they become an additional liability to the immediate emergency recovery period.

Disruption of any one of these critical facilities need not only be caused by major structural damage to the building system; in several cases less severe nonstructural damage has also been known to result in a complete breakdown of operations. For example, in Italy after the initial, main shock of the 1980 Irpinia earthquake, several hill towns remained without adequate water supply, sewerage facilities, and water treatment plants for weeks, and even months in some cases. After the 1989 Loma Prieta earthquake, the town of Watsonville was without a potable water supply for weeks after the ground shaking broke water and sewer distribution lines, mixing the two liquids together in the trenches where the pipes were located side by side. Several accounts of similar failures of critical urban support facilities following an earthquake are related below as quotes from available historic data.

SAN FRANCISCO, CALIFORNIA, 1906

> The state-of-the-art city-wide fire alarm system was knocked-out of action by the first shock wave. The writhing of the San Andreas fault not only broke telegraph lines and twisted streetcar tracks to stop all transit, it ruptured gas lines and water pipes. The gas fed flames from damaged fireplaces, flues and stovepipes, while the broken water mains rendered fire hydrants pressureless and firemen helpless.
>
> —From "The Day the Earth Shook" Dillon, 1985

MANAGUA, NICARAGUA, 1972

> The Managua central fire station, built in 1924 to withstand earthquake damage, was occupied by 20 firemen, 8 fire trucks, and 4 rescue ambulances. The main shock collapsed the second floor, crushing fire apparatus, killing 2 firemen and injuring others. The communication radio was destroyed and no emergency electric power was available. Fires soon began to break out in the city, where temporary hose lines were laid from the lake and pumps put into place because the local water system failed.
>
> —Bolt et al., 1977

SAN FERNANDO, CALIFORNIA, 1971

Most of the major structures in the heavily shaken area were medical facilities. Four major hospitals (Olive View, Veterans Administration, Holy Cross, and Pacoima Memorial Lutheran) were located within the radius of 9 miles of the epicenter. At the Veterans Hospital, some of the buildings that were built prior to 1933 collapsed. The other three hospitals, which were built within the last 12 years with earthquake resistance features, all suffered significant damage resulting in evacuation.

—Kessler, 1973a,b

CHILE EARTHQUAKE OF MARCH 1985

The City of San Antonio lost its water supply when the San Juan de Llolleo Pumping Station on the Maipo River suffered loss of electric power, severe settlement, and damage to buildings, well casings, equipment and pipelines. Much of the water in two tanks located on high ground in San Antonio was lost due to rupture of pipelines supplying them. San Antonio residents were being furnished water from Fire Department and rented tank trucks.

—Escalante, L., 1986

WHITTIER-NARROWS EARTHQUAKE OF OCTOBER 1987

[Regarding the operations of the Burbank Airport control tower after the earthquake,] an FAA supervisor decided to evacuate the tower since it was in an older building (60 years old) and utilize a mobile unit outside. Power to the tower was lost but emergency power did operate. One commercial plane ready to take off at the time of the earthquake was held with passengers on-board until operations resumed.

—Schiff, 1988

Another such failure happened after the Loma Prieta earthquake of October 17, 1989. Throughout San Francisco's Marina District, which was built on artificial man-made fill and reclaimed land from the 1915 Panama-Pacific International Exposition, numerous natural gas line failures occurred. Following the earthquake, the natural gas public utility supplier had to replace about 10 miles of gas lines. Total costs for this endeavor were estimated to reach as high as $100 million, indicating a linear cost of $10 million per mile.

Developments in Performance Standards for Critical Facilities

In evaluating the performance of these key critical urban systems following recent earthquakes, it has been recommended that the planning, design and construction of these emergency services facilities must focus on their ability to remain operable and functional after a major disaster.

In seeking improved performance standards for such essential, critical facilities, the key is found in the need to keep the facility not only "standing" but also "functional and operational" during and after an earthquake. In September 1975, the city of Los Angeles stated in its Seismic Safety Plan that "it is important for post-earthquake recovery that critical facilities such as police and fire stations, hospitals, dams and reservoirs, power facilities, and emergency communication systems remain operative after an earthquake."

CALIFORNIA HOSPITAL ACT OF 1972

As described earlier, the 1971 San Fernando earthquake plainly signaled that major acute-care hospitals located in urban areas of high seismic risk represent a particularly critical resource in response to an earthquake. It is important to repeat that in passing the Hospital Act, it was the specific intention that any new hospitals constructed in California "which must be completely functional to perform all necessary services to the public after a disaster, shall be designed and constructed to resist, insofar as practicable, the forces generated by earthquakes, gravity, and winds." For the first time in the history of California, performance standards for the design and construction of a building required that a specific facility remain functional and operational after an earthquake; this was in contrast to earlier state-mandated provisions, such as the Field Act (which addressed the safety of public schools), which included the limitation that a building survive an earthquake without injury to its occupants. The result was a new requirement in damage control in that not only the structural system, but also the architectural, mechanical, electrical, and life-support systems were expected to maintain their integrity and remain operational after an earthquake (see also Chapter 10).

In terms of reducing the potential of earthquake hazards, enactment of programs with higher seismic design standards for critical service facilities and land use planning restrictions in special study zones represent powerful actions. It is highly recommended for use in all areas of high seismic risk as a means of mitigating the effects of earthquakes relative to public health and safety.

EXISTING BUILDING STOCK

A primary reason that major cities remain extremely vulnerable to earthquakes is related to the fact that the majority of its existing building stock was constructed in past years when the knowledge and science of earthquake engineering may have not yet been fully developed. As indicated in Chapter 7, in the United States, building code standards are not retroactive despite the fact that it is not uncommon to find changes made to seismic provisions of the code after a severe, damaging earthquake.

Several changes were made to the Uniform Building Code (UBC) after the 1964 Alaska and 1971 San Fernando earthquakes. Even the 1985 Mexico earthquake was responsible for an addition of a fourth soil factor, S4, in the seismic provisions of the 1988 edition of the UBC.

As previously indicated in Chapter 7, on a national annual average in the United States new buildings added each year to urban environments account for about 2 percent of the existing building stock—the implication being that if a substantial change in building performance standards was introduced in one particular year, such as to ANSI A58.1 for example, one ostensibly could argue that technically speaking, the remaining 98 percent of buildings would not meet the new standards. Therefore, at any given time, the recognized vulnerability of the existing building stock in metropolitan centers, along with the existing infrastructure and critical emergency services, may change with the introduction of new performance standards.

Vulnerability of Existing Building Stock

The existing building stock in a metropolitan center is composed of many types of structures and building classes of diverse construction materials and age, including some designed before the promulgation of the provisions found in the first seismic code. Over the years, as a result of the pressures of increased population and urban growth, a typical city would have spread from its initial center and expanded into other areas having a myriad of diverse topographies: valleys, hills, mesas, canyons, flatlands, cliffs, coastlines, and plateaus. Thus, the seismic performance of these different building types and classes of construction, combined with the diverse behavior of geologic conditions under the structures in various areas, will not be uniform throughout the entire urban center. Anticipated building performance and damage patterns resulting from a severe earthquake in a metropolitan area will vary according to the number of building variables involved and the geologic characteristics underlaying their construction sites.

It must be realized that under such diverse conditions each building type will be vibrating at a different rate, depending on its size, mass, configuration, and class of construction. Over an entire area, with its different mix of buildings and dissimilar structures oriented in numerous directions, all moving at disparate rhythms, we will have a variety of performances. And as indicated in Chapter 7, since the relative safety of all buildings is not the same, we have to face the fact that the damageability patterns of all buildings are also not similar. In earthquakes, some buildings are more prone to damage than others, as shown in projected damageability curves.

With tens of hundreds of thousands of buildings in a complex urban setting, it is not hard to understand why so much damage may be realized after a severe earthquake. The 1985 Mexico City earthquake damaged approximately 14 percent of the total building stock in the central part of the city; 40 percent of the damaged buildings were corner buildings, either with two open fronts and two rigid sides or complex "flat-iron" shapes (see Figure 8-3).

In a congested, high-density urban center, the complexity of the problem also increases owing to the proximity of adjacent buildings, side by side on adjoining properties. This adjacency of structures in a downtown area or a central business district, where the pattern may be typified by high-rise, medium-rise, and low-rise structures of different types of construction, mass, and configuration, is the potential source of "dynamic pounding damage" between buildings. In extreme cases, effects of this "pounding" phenomenon have been known to lead to building collapse. (See Figures 8-4 and 8-5). The adjacency of a tall, flexible building next to a short, rigid one spells potential trouble, as this condition has resulted in severe damage to either building or to both. This proved to be a common cause of building damage in the 1985 Mexico earthquake. The combination of long-period ground motions and the long duration of ground shaking caused severe cases of dynamic pounding between adjacent buildings, which accounted for 42 percent of the building damage.

In terms of urban planning and design therefore, this situation presents an interesting question in addition to the problems of pounding damage themselves. How can pounding between adjoin-

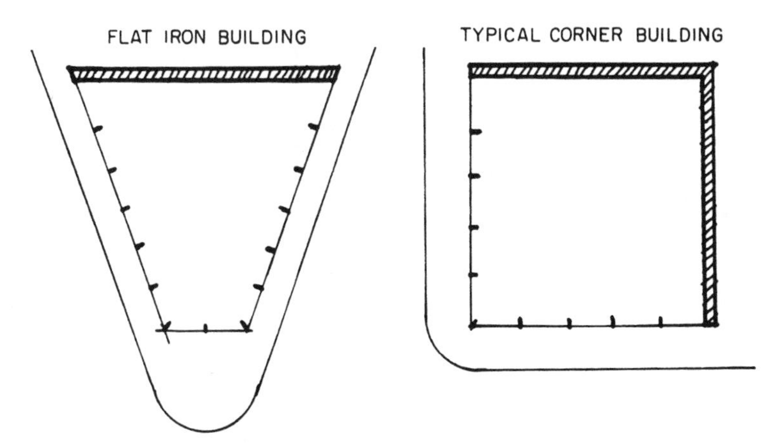

Figure 8-3 Plan views of flat-iron and corner building configurations, Mexico City, 1985.

Figure 8-4 Fourth-floor collapse of a five-story building owing to pounding of adjacent structure, 1985 Mexico City earthquake.

Figure 8-5 Result of pounding of two adjacent buildings, 1985 Mexico City earthquake.

ing structures be prevented when a new tall building is to be located next to a shorter existing structure? The simplest solution is to provide a seismic separation, or "seismic joint," between the two buildings. In areas of high land costs per square foot of property, such as in Tokyo for example, it should be recognized that the economic impact may be significant in that the developer of the new building must be prepared to accept a slightly smaller building rather than trying to cover every inch of the site.

Urban Vulnerability Analysis Studies

In undertaking earthquake preparedness planning for an entire urban center, one approach is to conduct a full-scale earthquake vulnerability analysis study of a selected metropolitan region to determine the impacts that a major seismic event might have on the immediate area. In the United States, 20 vulnerability study analyses have been completed through federal and state agencies since 1972 for various areas of high seismic risk around the country. Three more are being prepared for the San Diego/Tijuana (Mexico), Boston, and St. Louis metropolitan areas and are scheduled for completion in the near future.

The purpose of a vulnerability analysis study is to provide federal, state, and local government agencies with a rational basis for planning earthquake disaster relief and recovery operations in selected metropolitan areas where the repeat of a historic damaging earthquake has a high probability of occurring along a known active fault. For example, the very first vulnerability study to be completed in the United States was in 1972 for the San Francisco Bay Area by the Office of Emergency Preparedness, (Algermissen et al., 1972). Technically, the study for the area was based on the potential recurrence of earthquakes with magnitudes in the range of 6.0 to 8.3 on the San Andreas Fault and/or the Hayward Fault systems, not unlike the 1906 San Francisco or the 1836 East Bay Area earthquakes which occurred on these two faults, respectively.

The emphasis of a vulnerability analysis study is to present a worst-case earthquake scenario for planning purposes and to give a broad view of what may happen in an area in a severe earthquake. It does not propose that all damage patterns described will occur simultaneously during one earthquake, but rather presents a picture of damage "possibilities" that could occur during a credible seismic event. It does, however, signal and identify potential trouble spots that call for appropriate earthquake-preparedness activities. The purpose is to ascertain what should be done to improve public safety before rather than after the potentially damaging event. Such a study becomes a very useful tool for local governments to utilize in setting priorities for the reduction of losses and casualties in advance of a major earthquake.

In these studies, the vulnerability of all vital systems and physical facilities that make up the urban environment is documented and analyzed, including the following:

Police and fire stations	Medical facilities
Dams and reservoirs	Communication centers
Airports	Schools and colleges
Major freeways and highways	Railroads
Electric power lines	Natural gas lines
Fuel pipelines	Power plants
Sewer systems and lines	Water systems
Toxic storage facilities	Pumping stations
Harbors and ports	Rapid transit systems

These facilities are mapped on a base map indicating the geology of the study area so that the so-called "hot spots" of potential damage may be identified by matching geological hazardous areas with the location of critical facilities. For example, one conclusion that could be reached is that highways built on poor soils saturated with water along bay margins are expected to fail due to soil liquefaction under strong ground shaking. Another is that a massive earthquake-induced landslide could bury a water-pumping station and knock it out of service. Or that surface ground rupture along an earthquake fault line could sever a water supply distribution aqueduct.

Such postulated damage patterns, when correlated with one another, finally lead to the level of seismic vulnerability anticipated in the study area in the event of an earthquake. It is the responsibility of preparedness planners and local governments to deal with the problems identified and develop appropriate solutions to mitigate the hazards posed. In this way, earthquake vulnerability studies can help a community prepare for potential disasters and find ways for reducing their impact in urban centers. Architects must also learn how to interpret and use such studies as a useful tool in the development of large-scale urban planning and design projects.

FIRE FOLLOWING EARTHQUAKE: URBAN CONFLAGRATION

One of the great fears in the aftermath of a major earthquake is the threat of urban conflagration. The fear has been magnified over the years because of two dramatic historic events: the 1906 San Francisco earthquake and the Great Kanto (Tokyo), Japan, earthquake in 1923.

1906 San Francisco Earthquake and Conflagration

As a result of the severity of the 1906 San Francisco earthquake, many urban lifeline systems were ruptured or severed by intense ground shaking. The ensuing breaks in water, natural gas, and electric power lines set the scenario for the several fires that broke out immediately after the earthquake to be fanned into a three-day conflagration of such magnitude that it is said to have destroyed substantially more property and caused more casualties than the main shock. Lines supplying water to San Francisco from the main storage points were damaged or destroyed in vulnerable areas along the way to the city. In San Francisco itself, 300 breaks in the street water distribution pipe system were found and repaired.

The fire chief of San Francisco's fire department was killed during the earthquake. The city had no automatic fire alarm system in place. Fires broke out immediately after the earthquake, which struck at 5:12 A.M. By 8:00 A.M., about 50 fires were burning, although it is estimated that there were probably not more than a dozen that had started during the first half hour after the earthquake. In the general confusion that followed, the fires rapidly fanned out through the entire downtown district, spreading north and south along the diagonal Market Street area.

When the fire finally burned itself out, more than 75 percent of the city was destroyed. About 38,000 buildings had been burned out in an area covering 4 square miles, and some 300,000 residents were left homeless. Property losses at the time were estimated to be in the range of $350 to $400 million.

Great Kanto (Tokyo), Japan, Earthquake and Fire of 1923

A severe earthquake under Sagami Bay about 57 miles southeast of Tokyo occurred around noontime and resulted in a shift of the entire area to the southwest with a horizontal displacement of 15 ft and a vertical offset of about 6 ft. The earthquake, which was followed by a sea wave 36 ft high at Sagami Bay, destroyed more than half a million buildings and caused an urban conflagration that ran through the cities of Tokyo and Yokohama for more than three days. Nearly 70 percent of Tokyo and about 90 percent of Yokohama were destroyed by the earthquake and fire; there was $600 million in property damage and about 140,000 lives were lost. Fire damage was considerably more than damage caused by the earthquake.

At the time it was reported that there were 277 outbreaks of fire in Tokyo; 133 of these spread quickly through the urban areas fed by Japanese wood-frame construction, which was typical at the time of the earthquake. With broken water conduits and congested narrow streets compounding the problem, it was not surprising that the outbreaks of fire rapidly surged into an large-scale urban disaster. To this day, earthquake-preparedness programs in Japan are dominated by the expectation that urban conflagration is still a distinct threat.

Probabilities of Urban Conflagration

The potential of urban conflagration following an earthquake depends on at least four elements: (1) a trigger mechanism such as an earthquake or other type of disaster; (2) the presence of considerable amounts of combustible material, or "fire load," in the impacted area; (3) the amount and performance of fire-response personnel, equipment, and facilities, including the availability of an adequate water supply; and (4) positive or negative weather conditions, including wind velocity. If the negative aspects of all four elements are present at the time, the outbreak of large-scale urban fires is highly likely.

Planning Countermeasures

There are several countermeasures from the urban planning point of view that are effective against fire spread. One of the most important is keyed to the amount of combustible materials in the area. From the planning point of view, therefore, it makes sense, as is commonly required under current fire codes, to limit the use of combustible materials in construction in high-density congested urban centers. While more flammable materials of construction were used at the turn of the century, most metropolitan areas in the United States now restrict the amount of wood construction and use of other combustible materials. However, since the use of synthetic materials has increased to a critical point today (being that they are used in carpeting, decorative elements, bathroom/kitchen equipment, architectural detailing and trim pieces, upholstery, etc.) and our buildings store many more toxic chemicals than ever before, the caution flag must be raised.

Another planning element that can help prevent fire spread concerns the width of streets. A street can serve as a natural firebreak if it is wide enough. The Tokyo/Yokohama metropolitan areas in 1923 were characterized by narrow streets and tightly congested alleys. Not only was it impossible to move fire-fighting equipment through the streets after the earthquake, but there were no natural firebreaks in the areas except for riverbeds. In the 1906 San Francisco earthquake, the dynamiting of buildings and deliberate burning of wood structures as backfires along the wider Van Ness Avenue (125 ft) finally formed a firebreak sufficient to stop the surging fire, even though reports indicate that the fire did reach as far as Octavia Street and Golden Gate Avenue to the west. Clearly, in the postearthquake planning and reconstruction of urban centers (and as was done in the reconstruction of the city of

Tangshan, China; see Chapter 9), designing thoroughfares to be wide enough significantly contributes to safety in areas of high seismic risk.

SOCIAL AND ECONOMIC CONCERNS FOR PUBLIC HEALTH AND SAFETY

In presenting the social and economic effects of earthquakes, the statistics that capture the front pages are usually related to human casualties and dollar losses. Another aspect of the impact of earthquakes that must be taken into consideration from the urban planning and design point of view deals with the problem of the number of people rendered homeless by damage resulting from the main earthquake and/or evacuated in anticipation of severe aftershocks. A secondary social impact is the loss of educational facilities when schools are damaged enough to warrant long-term closure.

Earthquake-Induced Homelessness

The magnitude of this problem can be overwhelming: for example, nearly 2 percent of Mexico City's population, approximately 300,000 people, were rendered homeless in the 1985 earthquake. In Armenia's earthquake in 1988, about 500,000 lost their living units and became homeless. In an earthquake's immediate aftermath, this type of problem has an enormous social impact, since there already is a great deal of confusion and local government is busy wrestling with many other aspects of the crisis throughout the entire impacted area. Overwhelming numbers of people made homeless also pose a huge economic problem. Not only must shelter be found for them, but they must also be fed.

Fortunately, in Mexico City (and in other earthquake sites as well), many were able to rely on their extended family and close friends for help with food and shelter. According to Richard Eisner (1988), director of the Bay Area Earthquake Preparedness Project (BAREPP) in the San Francisco Bay Area, it was estimated at the time that roughly 50 percent of the homeless in Mexico City somehow managed to find their own shelter; some used underutilized government emergency shelter facilities and others camped on vacant property or in the open in parks near their destroyed homes so that "they could maintain neighborhood and community ties." During the recovery phase that follows such situations, the provision of adequate housing for the homeless represents a critical urban planning and design problem. This aspect of the homeless problem and its phased solutions are explored in greater detail in Chapter 9.

Loss of Educational Facilities

Schools are a very important resource in any culture for the education of future generations. The young are not as resourceful as adults, and experiencing dislocation owing to damaged school buildings, death and injury of classmates, and loss of educational continuity can be traumatic. This social issue is also an emotional one, since the safety and security of a defenseless child are considered inviolate.

It is for this reason that California passed the Field Act in 1934 following the 1933 Long Beach earthquake. Fortunately, schools were not in session when the earthquake struck, so no casualties occurred. But the specter of potential disaster was vividly imprinted on the public's mind, enough so that it ensured passage of state legislation that required schools to be classified as a special building type, to be designed and built under higher earthquake performance standards than other structures. The Field Act has served the state very well, for despite the many new school facilities that have been built and all the earthquakes in California since 1934, public school buildings have performed extremely satisfactorily owing to their stricter design and construction. The Field Act represents a model that other countries and states should adopt for the design and construction of public schools. For example, the 1985 Mexico earthquake severely damaged many schools, and there were several spectacular collapses.

Economic Impact

Damage losses are typically quoted in dollars lost at the time of the actual earthquake. However, this is only part of the story. The financial toll may continue for a long period of time. It is difficult to measure the total economic impact until everything returns to normal and full and accurate assessments can be made. But many other economic losses are commonly not reflected in this total figure, including loss of wages and productive income because of commercial inactivity, loss of tourism when travelers avoid the stricken region, and loss of exports to other countries (an effect of which may be trade imbalances) because of curtailed industrial production. A severe earthquake may easily result in a noticeable loss in the gross national product of a developing country, since the productive capacity of a vital region may be affected until recovery is complete.

In many countries tourism is a vital industry because of the hard currencies brought in by foreign visitors. Mexico is no exception. Because of the heavy damage sustained in Mexico City in 1985, many tourist hotels were destroyed and entirely disappeared from the map: Hotel Regis, Hotel del Prado, Hotel Bamer, Hotel Alameda, Hotel Continental, Hotel del Paseo, and Hotel Monejo, to mention a few. Others were severely damaged and rendered inoperational for a year or more. Furthermore, tourists frightened by the negative publicity of the awesome earthquake avoided Mexico City for more than a year. It took about two years for the city to recover its popularity as a tourist destination, and this resulted in a large loss of income for workers in Mexico's tourist-related industries.

Exactly the same thing happened in San Francisco after the October 1989 Loma Prieta earthquake. Many tourist hotels, though relatively undamaged, found themselves with low occupancy levels as vacationers and conventioneers left as quickly as possible. A substantial number of conventions scheduled for the months immediately following the earthquake were even canceled outright. Two months after the earthquake, the mayor of San Francisco flew to various states on the East Coast to convince potential travelers that his city had recovered from the earthquake, was a safe place to be, and could handle any accommodations for conventions and annual meetings.

Probably the severest economic impact experienced after a major earthquake catastrophe is the cost of recovery and reconstruction, which is the principal subject of the next chapter.

CHAPTER

9 RECOVERY AND RECONSTRUCTION

Following a major damaging seismic event in a metropolitan area, the recovery process begins. This recovery process includes an immediate short-term emergency relief and recovery phase followed by a longer-term reconstruction period. Each phase contains distinctive characteristics, different objectives, and as implied, covers diverse time periods. The main thrust of the emergency recovery phase focuses on saving lives, administering medical aid to the injured, providing emergency shelters, distributing food and water, reestablishing communication and transportation systems, and supporting economic productivity. The longer-term reconstruction phase emphasizes rebuilding of the urban environment and all its support services.

In this chapter, the focus will be on those recovery and reconstruction activities in which architects and physical planners have a direct role to play. It is not that other aspects of the recovery phase are considered unimportant, but rather that they would be better handled by others with a greater expertise in the topic area. Accordingly, the reconstruction of lifelines, political decision-making policy, and other nondesign considerations—important as they may be—will not be covered in specific detail.

A review of historic data indicates that the focal depth of an earthquake is a significant ingredient that influences the degree of damage and number of casualties. This is notably true when the earthquake's focus is located directly beneath the impacted urban center, the so-called "bull's-eye earthquake."

Recently, research work has been conducted to determine if it is possible to supplement the Richter magnitude scale with a sort of composite magnitude to include a measure of the differences between deep-focus and shallow earthquakes. Such a composite measure would complement the Richter magnitude scale, which was never meant to be an indicator of the earthquake's depth of focus. Table 9-1 presents the number of deaths resulting from three

TABLE 9-1 Comparative Analysis of Deaths Resulting From Three Earthquakes of Diverse Magnitudes and Focal Depths

Earthquake	Magnitude	Depth of Focus (km)	Number of Deaths
Seattle, 1965	7.1	50	7
Managua, 1972	6.5	5	10,000
San Salvador, 1986	5.4	5	1,500

Source: Algermissen (1989).

selected earthquakes of diverse magnitudes that occurred more or less directly beneath the cities indicated, (Algermissen, 1989).

It is interesting to note in Table 9-1 that the largest earthquake, Seattle's, resulted in the smallest loss of life. At first glance this seems strange, at least until we examine the focal depth for each of the three events indicated. Although this measure is one among many others influencing life loss (the others being vulnerability of existing building stock, the direction and characteristics of fault rupture, geological setting, magnitude of the earthquake, location of the epicenter, demographics of the urban area, etc.), it does appear to be a significant one.

In contrast to these three earthquakes, and as proof that it is not possible to deal in generalities when discussing seismic safety issues, the 1985 Mexico event resulted in the heaviest damage and life loss in Mexico City, owing to long-period ground motions and lakebed soils amplification 250 km away from the focal point and epicenter, which were located on the coast. Admittedly, this comparison may be questionable, because Mexico City has a population near 20 million, while the three cities in Table 9-1 contained much lesser populations. Nonetheless, important lessons are to be learned from all four earthquakes.

IMMEDIATE EMERGENCY RECOVERY PHASE

This part of the postearthquake recovery period occurs after the ground shaking stops, buildings stop oscillating, and an eerie reaction of numbing shock spreads throughout the city. It is not unusual at this point to find much dust settling in the severely damaged parts of the urban center, depending on the number of buildings that have collapsed. For example, in the 1988 Armenia earthquake, 10 surveillance helicopters dispatched to evaluate the degree of damage were unable to obtain a total assessment of the destruction in the region because of dust clouds that hindered accurate observation over cities in the area such as Spitak (the earthquake's epicenter), Leninakan, and Kirovakan. Many victims extricated from damaged buildings also indicated respiratory problems caused by large amounts of dust in the area where they were trapped.

It is also during this phase that many frantic emergency activities begin simultaneously, as everyone who can respond to the event starts taking relief actions. One of the very first reactions is to ask what happened and then attempt to determine exactly where it happened. Provided that the instruments were working properly (the seismic station located in Leninakan, for example, was destroyed in the 1988 Armenia earthquake), probably the only hard evidence available at this point is a seismograph record of an event that was recorded at one of the official seismograph stations. This record hopefully will provide data on the magnitude and regional location of the earthquake, but unfortunately will not give any clue as to the extent of damage and casualties in a specific location or community.

Even in the immediate affected zone itself, particularly if it covers a large area, it is quite common at this point not to know the full extent of damage nor to be able to pinpoint exactly where the worst damage or collapses are. More than likely, communication networks have been disrupted and electric power lost, so it is not a simple matter of picking up the phone and calling information services or emergency response offices with such inquiries; anyway, at this point they probably won't have this information either.

After the 1976 Tangshan earthquake, due to disrupted communication lines, two days passed before the Chinese government offices in Beijing understood the full extent of damage in the city of Tangshan. The only way that Beijing was able to receive a comprehensive picture of what had happened was to dispatch an army jeep cross-country to Tangshan to find out. Because of collapsed bridges and highway overpasses, the jeep had to make several detours before its crew could reach the heavily damaged city and make a detailed report. During the two-day period that it took the jeep to reach Tangshan, there had been many self-directed activities already begun inside the devastated city. In the 1988 Armenia earthquake, 20 hours passed before complete assessments of damage patterns were ready for review.

Care for Casualties and Search and Rescue Activities

Immediately after an earthquake, relative confusion reigns, yet several ad hoc self-directed activities and official local government actions take place, provided the facilities of response agencies remain operational. However, depending on the earthquake's severity, initial recovery efforts may be hampered by roads having been rendered impassable, access streets that have been blocked by fallen debris, and the absence of traffic control during the first hours.

Early actions and activities may include (1) emergency care for casualties; (2) immediate collateral hazards mitigation such as fire following earthquake, potential downstream flooding following dam failure, and explosion of toxic materials and chemicals; (3) debris removal for access to damaged areas and for traffic control; (4) assistance in evacuation of buildings and critical areas; (5) reestablishment of communications; (6) damage and usability assessment of buildings; and (7) search and rescue—and not necessarily in that order. At this early stage it is advisable to develop a priority list of what to do first.

An example of a first priority item in the October 1989 Loma Prieta earthquake in northern California was the explosion and fire in San Francisco's Marina District, which were caused by ruptured natural gas lines; the situation required immediate attention by the fire department. As the Marina District consists of several blocks of wood-frame dwellings, the fire quickly spread to adjacent buildings, and the specter of a repeat of the 1906 San Francisco earthquake and fire was uppermost in everyone's mind. Luckily, the fire department was able to respond rapidly, and although water supply was limited, hoses were quickly run to the nearby bay waters and the fire was totally extinguished within 19 hours (see Figure 12-1).

Another highest priority, by necessity, is to provide care for the injured and to see that no further casualties are sustained. Assistance in the evacuation of major buildings or hazardous areas of the city may be necessary, depending on the levels of damage. Emergency medical treatment is required, and depending on the condition of medical facilities following the event, field hospitals may be established in appropriate open spaces found near or adjacent to the stricken city. After three of the four hospitals were destroyed in Leninakan in the 1988 Armenia earthquake, a military field hospital was set-up after the second day of the earthquake near the town of Kirovakan; it performed about 1250 operations in a 2 and 1/2-week period.

Of all the actions taken in this regard, perhaps the most dramatic is the work of search and rescue teams in locating and extracting live persons trapped in collapsed buildings. This task means working against time, as aftershocks may lead to the further collapse of structures damaged by the main earthquake and the death of injured and uninjured victims still trapped in damaged buildings. Victims trapped by building collapse and rescued within the first 24 to 36 hours are given the best odds for survival; their chances of being rescued diminish beyond 72 to 96 hours. For this reason, search and rescue teams often refer to the first 24 hours as the "golden twenty-four." However, it is interesting to note that some victims have been extricated alive from damaged buildings even much later than the first 24 to 36 hours. After the 1990 Philippines earthquake, two survivors were rescued and extracted alive from the rubble of a collapsed hotel building 11 days after the 7.7 magnitude earthquake occurred. Some rescue problems encountered during the chaotic period immediately following the earthquake include inadequate transportation, lack of communications systems, scarcity of drinking water, hospital capacities overrun by casualties, sparseness of available medical supplies, shortage of electric power, and significant levels of dust.

The 1988 Armenia earthquake provides an excellent case study model for search and rescue actions. A medical investigation team from the University of Pittsburgh reported the number of trapped victims extricated and rescued from damaged buildings in the impacted area on a daily basis, including the number of persons evacuated from the immediate area to other cities in Armenia and other republics (see Table 9-2).

There are reported cases of victims trapped for long periods who have been removed alive from a collapsed building only to die later of "crush syndrome" and shock. According to Dr. Eric Noji of Johns Hopkins University in Baltimore in reporting on his experiences after the 1988 Armenia earthquake,

> Crush syndrome is a condition that develops secondary to significant muscle damage. Injured muscles release intracellular substances into the general blood circulation that are extremely toxic to several vital organs, particularly the heart and kidneys. The treatment for crush syndrome is prompt administration of intravenous fluids that prevent the buildup of these substances to toxic levels as well as keeping the kidneys well-perfused and functional. Unfortunately, as noted above, very few persons who have been successfully extricated from the debris received any advanced medical care on-site or in transport to the hospital, including the administration of intravenous fluids. By the time they reached the hospital, many of these patients were severely dehydrated or in hemorrhagic shock. . . . There were several anecdotal reports of patients who seemed to be recovering after several days in the hospital who expired suddenly for no apparent reason. It is postulated that they may have died of fatal heart arrhythmias secondary to potassium being released from damaged muscles (Noji, 1989).

In reviewing Table 9-2, it should be noted that heavy equipment was not available for search and rescue teams until days three to

TABLE 9-2 Trapped Victims Extricated, Rescued and Evacuated, 1988 Armenia Earthquake[a]

Category	Day 1	Day 2	Day 3	Day 4	Day 5	Days 6–12	Days 13–19	Total
Total extricated	4.3	9.6	8.2	6.4	4.4	8.2	0.4	41.5
Extricated alive	1.4	1.7	4.8	5.7	1.8	0.15	0.001	15.5
Total evacuated		2.5	0.1	1.7	4.1	59.6	36.4	104.4
Total evacuated to other republics					1.3	35.0	29.2	65.5

Source: Klain et al. (1989).

[a]Rounded figures and totals given in thousands.

four. Until then, the initial relief response was generally led by relatives, neighbors, and additional ad hoc groups. The major causes of life loss in collapsed structures appear to have been crush syndrome, hypovolemic shock, cranial trauma, and multiple injuries. Search and rescue teams that locate persons trapped in the rubble of collapsed buildings are composed of a small group of trained and experienced professionals; they have access to techniques and specialized equipment for locating and extracting victims, such as "sniffer" dogs, sensitive listening devices, hydraulic jacks, and heavy debris removing tools, and they are able to work in small spaces. In some cases, the small space in which a survivor may be trapped may have been formed by a nonstructural building element, such as a sturdy metal table or desk, holding collapsed floors or beams sufficiently apart long enough for the victim to be saved by the search and rescue team (see Figure 9-1).

The major medical problems most often found by rescue teams include crush injuries, fractures, shock, dust inhalation, infections, open bleeding lacerations, and injuries requiring amputation. Despite these problems, recent success rates in locating and saving trapped victims through the use of urban heavy rescue teams have been significant. As urbanization continues at a rapid pace throughout the world, search and rescue teams will assume greater importance in saving lives after an earthquake in which major damage patterns occurred. Urban heavy rescue procedures by their very nature contain such a high degree of emotional visibility that their value in attempting to save even the very last victim cannot be ignored by public policy decision makers.

Figure 9-1 Collapsed school building, 1985 Mexico City earthquake.

Assessment of Building Damage

One task that must be accomplished immediately after a severe earthquake is a comprehensive assessment of building damage. Residents of the stricken area must know in no uncertain terms which buildings are considered safe to reoccupy and which are not. The first questions commonly asked by local residents and building owners, governmental or otherwise, relates to this aspect of the recovery phase. Typical questions asked are exemplified by the three following examples: (1) Is my house (apartment complex, office building, or whatever) safe to enter? (2) Will it survive the next aftershock or is it a safety hazard? (3) Is my damaged building repairable or is it targeted to be demolished for public health and safety reasons?

Such critical questions cannot be answered haphazardly nor by a layperson. In addition, they must be answered quickly and accurately as soon as possible if the initial stages of recovery plans are to start smoothly and on an objective, realistic basis.

Only building design professionals, architects or engineers, and building department officials have the appropriate credentials to answer such crucial questions. Furthermore, experience clearly shows that in a major disaster the capacity for response by local professionals and authorities is completely overwhelmed during the first few days immediately following the event, so outside experts have to be brought into the stricken area to supplement and assist local representatives in the damage assessment task. In Naples after the 1980 Campania-Basilicata earthquake, a technical cadre of 2000 were required to complete an inspection of buildings to determine their safety. Unfortunately, following a major aftershock three months later that caused additional damage, they found themselves repeating the assessment.

Architects can play an important role, along with others, in this essential task of assessing building damage rapidly so that local authorities may obtain a clear picture of the extent of damage in the area and the status of their community before proceeding with reconstruction plans. Unfortunately, there is no shortcut to the assessment process. A very time-consuming building-by-building inspection in the field must be done and official data sheets compiled.

Besides obtaining a comprehensive picture and an official record of the overall damage to an urban center after an earthquake, another fundamental purpose of a rapid assessment of damaged buildings is to prevent further casualties by carefully identifying buildings weakened by the earthquake and subject to additional damage, or collapse, in subsequent aftershocks. As such, it can also serve as a data base to assist in determining the number of temporary housing units needed during the period immediately following the earthquake.

TABLE 9-3 Damage Assessment and Usability Index of Building Damage

Category	Color Posted	Usability Index	Damage Level
1	Green	Usable	None to slight
2	Yellow	Temporarily usable	Moderate to heavy
3	Red	Unusable	Severe to partial or total collapse

Source: Rojahn and Reitherman (1989), ATC-20, Applied Technology Council.

In identifying levels of damage in buildings, the inspection process is also used to post buildings as to their relative safety. The practice widely used in most parts of the world is to utilize a system of marking buildings by color-coding or symbols signifying a safety scale. Table 9-3 explains the meaning of a three color-coded system typically used in the posting of damaged building based on the criteria developed by the Applied Technology Council (ATC) of California.

Other systems used by field inspectors are to mark each structure with a symbol, such as a large red "X" or "O" painted directly on the building, to differentiate those damaged buildings that were identified as repairable from those judged beyond repair and to be demolished. After the 1980 earthquake in southern Italy, slightly damaged buildings were painted with the word "AGIBILE" on the exterior wall near the main entrance to indicate that they were repairable and could continue to be occupied by the owners or tenants while repairs were being made. Buildings marked "NON AGIBILE" were impractical or infeasible for repair and occupancy (see Figure 9-2).

Under the system developed by the ATC through the Bay Area Regional Earthquake Preparedness Project (BAREPP) for use in California, severely damaged buildings are to be posted by red placards that say "This structure has been seriously damaged and is unsafe. Do not enter. Entry may result in death or injury." Bold letters on the red placard also read "UNSAFE, DO NOT ENTER OR OCCUPY." At the other "safe" end of the two-scale system, a building that passes inspection is posted with a green placard that reads "INSPECTED, NO RESTRICTION ON USE OR OCCUPANCY" and includes the following commentary: "This structure has been inspected (as indicated below) and no apparent structural hazard has been found. Report any unsafe conditions to local authorities; reinspection may be required." It is reassuring to note that the colors red and green have the same connotation as the colors recommended by a UN/Balkan study group. Figure 9-3 shows the ATC-20

Figure 9-2 Severely damaged building posted unsafe for occupancy, 1980 Campania-Basilicata, Italy, earthquake. *Source:* Mader and Lagorio (1987).

Rapid Evaluation Safety Assessment Form developed in 1989 by the ATC for BAREPP. This form was officially used for the first time in the posting of buildings after the October 1989 Loma Prieta earthquake (see Chapter 12).

No matter what system is used, the assessment forms must be direct, simple, easy to read, and designed for ease of use by the field inspectors. Careful development of any damage assessment procedure is necessary to obtain uniformity in the rating of building damage. To be successfully operative and effective, the system used should be such that if two separate individuals examine the same building, the results should essentially produce the same recommendations relative to its safety level, posting, and usability index.

To be effectual, the field inspectors must be quite familiar with local construction practice and technically knowledgeable of typical building damage patterns. When there is a need to call on field inspectors from other regions, it is extremely advisable to have a short, general training session followed by technical briefing sessions at the end of each day for information exchange and correlation of data. It is apparent that architects have a role to play in this building inspection period, especially where wood-frame construction and other small structures are concerned. However, it is essential that the architect be part of the interdisciplinary inspection team with other members from structural engineering and the local department of public works or building department, which is authorized to do the actual posting of buildings.

Block__________ Parcel No.__________

ATC-20 Rapid Evaluation Safety Assessment Form

BUILDING DESCRIPTION:
Name: ______________________________________
____
Address: ______________________________________
____
No. of stories: __________
Basement: Yes ☐ No ☐ Unknown ☐

Primary Occupancy: Dwelling ☐
Other Residential ☐ Commercial ☐ Office ☐
Industrial ☐ Public Assembly ☐ School ☐
Government ☐ Emer. Serv. ☐ Historic ☐
Other______________________________________

OVERALL RATING: *(Check One)*
INSPECTED (Green) ☐
___ Exterior only
___ Exterior and Interior
LIMITED ENTRY (Yellow) ☐
UNSAFE (Red) ☐

INSPECTOR:
Inspector ID______________________
Affiliation ______________________

INSPECTION DATE:
Mo/day/year______________________
Time ______________________ am pm

Instructions: Review structure for the conditions listed below. A "yes" answer to 1, 2, 3, or 5 is grounds for posting entire structure UNSAFE. If more review is needed, post LIMITED ENTRY. A "yes" answer to 4 requires posting AREA UNSAFE and/or barricading around the hazard. Hazards such as a toxic spill or an asbestos release are covered by 6 and are to be posted and/or barricaded to indicate AREA UNSAFE.

Condition	*Yes*	*No*	*More Review Needed*
1. Collapse, partial collapse, or building off foundation	☐	☐	☐
2. Building or story noticeably leaning	☐	☐	☐
3. Severe racking of walls, obvious severe damage and distress	☐	☐	☐
4. Chimney, parapet or other falling hazard	☐	☐	☐
5. Severe ground or slope movement present	☐	☐	☐
6. Other hazard present	☐	☐	☐

Recommendations:
☐ No further action required
☐ Detailed Evaluation required (circle one) Structural Geotechnical Other__________
☐ Barricades needed in the following areas: ______________________________
____
☐ Other: ______________________________

Posted at this Assessment: ☐ Yes ☐ No

Comments:______________________________________
____
____
____
____

Figure 9-3 ATC-20 form used to post a building safe for occupancy. *Source:* Applied Technology Council.

Loss of Housing Units

A general statistic in measuring damage patterns due to damaging seismic events on a worldwide basis is that 30 percent of all earthquake damage involves that to residential structures. Of these, any dwelling suffering 50 percent or greater loss is considered to be uninhabitable, since the utilities are usually inoperative, the building system has experienced substantial structural damage, and nonstructural operational elements are typically impaired (doors and windows won't open or close, light fixtures can't be turned on, mechanical/bathroom fixtures/kitchen appliances have been rendered inactive, etc.).

Accordingly, it is not uncommon to find a large number of victims rendered homeless who need temporary, emergency shelters immediately after the earthquake. As indicated in Chapter 8, there were 300,000 homeless in Mexico City after the 1985 Mexico earthquake, and 500,000 homeless after the 1988 Armenia earthquake. In the United States, a vulnerability study completed in 1985 indicates that about 270,000 people in the greater St. Louis area would need emergency shelter because of damaged or destroyed homes if a magnitude 7.6 earthquake occurred on the New Madrid Fault Zone. The estimate is based on the historic record documented during 1811–1812, when three earthquakes above an 8.0 magnitude occurred on the New Madrid Fault.

In response to needs of the homeless during the postearthquake recovery phase, replacement of damaged housing units takes place in three distinct stages:

Stage 1. *Emergency Shelter:* Immediate emergency period following the earthquake.

Stage 2. *Temporary Housing:* Intermediate recovery period following the earthquake.

Stage 3. *Permanent Housing:* Basic reconstruction period.

EMERGENCY SHELTERS

During Stage 1, the objective is to provide literally anything immediately available that can serve as a basic shelter during the chaotic time in which response groups are still marshaling their forces and developing plans of attack for recovery. After the 1985 Mexico earthquake, tents were used extensively for this purpose throughout Mexico City while the government prepared other official shelters. (See Figure 9-4.) In addition, people used their extended family and networks of friends as much as possible. In the severely damaged Tlatelolco residential complex, where sections of a large multistory housing block of units collapsed, displaced residents lived and slept in tents and open lawn areas in an atmosphere of outdoor camping rather than be relocated to government shelters.

Figure 9-4 Tents used as emergency shelters, 1985 Mexico City earthquake.

After the 1980 Campania-Basilicata earthquake in Italy, the government sent hundreds of large metal shipping containers to Naples and railroad cars to the hill towns in the stricken areas for use as emergency housing units. (See Figure 9-5.)

In the United States, right after the May 2, 1983, Coalinga earthquake, where the majority (1900 units of 2700) of housing stock was severely damaged or destroyed (see Figure 9-6), emergency temporary shelter for residents was very informal; individuals slept in camping trailers parked in front of their damaged homes or on

Figure 9-5 Metal container units used as emergency Shelters, 1980 Campania-Basilicata, Italy, earthquake. *Source:* Mader and Lagorio (1987).

Figure 9-6 Severely damaged residence, 1983 Coalinga, California, earthquake.

their front lawns in tents or just in sleeping bags, as the weather at that time of year was benign. A few days after the earthquake, a large circus-style tent was assembled at a local community college for a mass care shelter and food facility for those without any type of emergency shelter.

The period of time during which emergency shelters are used varies according to the severity of earthquake losses and the capability and capacity of government response. It could be anywhere from a week to one or two months at the most before more substantial housing units are made available on a temporary basis. In one severe case, it was a matter of four to six months before the next level of appropriate housing units were made available.

In China after the 1976 Tangshan earthquake, the approach to emergency housing was to rebuild residential units in the same areas as rapidly as possible, using indigenous materials and masonry rubble of the devastated buildings. (See Figure 9-7.) Dwelling units were built in which whole families could live temporarily without the need for another move until new permanent housing was completed. This policy on the part of the central government preserved local social structures and allowed appropriate time for careful planning of the new city. Some residents were evacuated to other areas. Parks in Beijing and Tianjin provided refuge for 300,000 after the earthquake. Open space at the Tangshan airport served as the main medical station.

Emergency shelters typically are not expected to necessarily have water hookups, electric power, sewerage systems, or kitchen/cooking facilities. Their primary purpose is to provide a safe sleep-

Figure 9-7 Temporary housing constructed with brick debris, 1976 Tangshan, China, earthquake. *Source:* Ministry of Construction, People's Republic of China.

ing space in a somewhat sheltered and comfortable environment. Obviously, the more benign the climate at time of their use, the better. In severe climates it is necessary to replace those crude, emergency shelters with more environmentally protective units as soon as possible, or arrange for the evacuation of the large numbers of homeless to regions with less severe climates. However, the latter may be difficult to achieve, as experience indicates that residents prefer to remain near their damaged or destroyed homes in order to maintain neighborhood and community ties and be part of the reconstruction of their own areas, rather than be moved to another location outside the urban region.

Both in the 1980 earthquake in southern Italy and the 1988 earthquake in Armenia, a large part of the temporary housing problem was solved by evacuating residents to other regions not affected by the seismic events. In Italy, since the 1980 earthquake occurred in late November, when most tourists had left, tourist hotels in coastal cities were used as temporary housing for those evacuated from the damaged hill-town areas. There was a great deal of resistance to this move, but since virtually no type of housing existed in many of the destroyed hill towns, the displaced begrudgingly agreed to it.

Similar evacuation plans were used after the 1988 Armenia earthquake, which occurred in the severe winter temperatures of December and in which the entire existing housing stock was practically wiped out in many areas. Obviously, residents could not survive in the freezing weather without shelter and heat. Since no shelters existed and the homeless numbered about 500,000, the central government concluded that while waiting for new permanent housing to be constructed in Armenia, the only practical solution was to move as many residents as possible to more hospitable climates where some type of housing existed. Over 170,000 residents were evacuated from the area impacted by the 1988 earthquake, 65,500 to other republics of the USSR. As in Italy, the economic base of the severely damaged Armenian cities and towns was principally agricultural, and many residents had emotional attachments to the land spanning generations. Again, there was initial resistance to the move, but eventually it had to be accepted.

TEMPORARY HOUSING

Hopefully, emergency shelters in place immediately after a disaster are replaced as soon as possible by more substantial housing units to be used during the initial phases of the reconstruction program that follows. The second stage relies on the availability and utilization of temporary but more consequential housing unit types.

Typically, these second-stage housing units are expected to be used for a considerable period of time, for the longer-term rebuilding/reconstruction process may take several years. Because of this potentially sustained period of occupancy, the question is always raised as to the danger of a temporary facility inexplicably becoming a permanent one. Obviously this has proven to be the case in many instances and needs to be guarded against, particularly when economic pressures become acute.

In this regard, Mexico City was an exception in that a special emphasis was placed on public housing as a high-priority item during the reconstruction phase; one development of 34,500 permanent units was built within the first year of the 1985 earthquake using construction techniques that fostered owner participation. Similarly, after the 1988 Armenia earthquake in the USSR, government authorities in Moscow gave high priority to the construction of new housing units, a few of which were already completed as demonstration projects within 10 months of the earthquake. In contrast to the Mexico City experience, however, these new housing units were constructed without resident participation, probably due to the fact that entire sections of the cities, such as Spitak and Leninakan, were generally leveled by the earthquake. (See Figure 8-1.)

Temporary housing, like emergency shelter solutions, comes in many shapes and sizes. Even before the temporary units are available, one of the first decisions is where they are to be placed. Because these units are more substantial than the emergency shelter units and may require some type of power, water, and/or sewerage system hookup, site location becomes a very important. For example, in the case of one of the southern Italian hill towns heavily damaged by the November 23, 1980 earthquake, about 30 temporary housing units arrived amid a political debate in the mayor's office as to which site was to be used. Finally, in desperation, instructions were issued for their delivery to and assembly on a specific open-space location, only later discovered to be in the middle of an unstable massive landslide area. Fortunately, after more heated debate, the temporary units were moved again, this time to a more suitable site.

In many of these Italian hillside areas, the only large, level open space appropriate for the location of temporary housing units turned out to be the local soccer field. Thus, in many hill towns this public athletic field became the site of temporary units while plans were being made for the reconstruction of damaged housing and/or the siting and design of new units. (See Figure 9-8.) Another

Figure 9-8 Prefabricated temporary housing units placed in open field, 1980 Campania-Basilicata, Italy, earthquake. *Source:* Mader and Lagorio (1987).

convenient, relatively flat open space available for the rapid placement of temporary housing units in the area was found to be the sides of paved major streets outside of the main center of town. "Ruolottes," or mobile housing units mounted on wheels, were lined up on both sides of wide thoroughfares. (See Figure 9-9.)

After the 1985 earthquake in Mexico City, temporary housing units, first of heavy kraft board and later of corrugated sheet metal, were assembled in streets closed off to traffic and adjacent to construction sites proposed for the permanent housing program that followed. This allowed the displaced residents to maintain neighborhood cohesiveness and actively participate in the rehabilitation and repair of their damaged homes and/or the construction of new residential units.

In addition to serving basic living needs, temporary residential units should be as light weight as possible and easily transportable, demountable, and movable. In most cases these programmatic attributes are easily translated into the design of appropriate units, which are typically prefabricated. Here again, the architect will have a role to play in their design and construction details.

In 1983 the Federal Emergency Management Agency (FEMA) brought in approximately 200 temporary housing units in the form of house trailers all the way from Utah to Coalinga, California. They were stored at a vacant railroad right-of-way south of the town's

Figure 9-9 Camping trailers used as temporary housing units, 1980 Campania-Basilicata, Italy, earthquake. *Source:* Mader and Lagorio (1987).

Figure 9-10 Portable prefabricated units used as temporary housing in open field, 1983 Coalinga earthquake.

center until they could be located on two available sites previously identified by the Coalinga General Plan as being suitable for the future development of two mobile-home parks. (See Figure 9-10.)

Temporary housing is only meant to serve residents during a holding period while new permanent housing is being constructed. Depending on the culture, public participation in the planning, design, and construction process may or may not occur. Typically, countries with central governments do not encourage resident participation in the decision process. However, experience has shown that the entire process works more smoothly and that there is greater acceptance of the new housing units when public participation is involved. In such cases, it has also been found that there are less objections to moving to new housing.

NEW PERMANENT HOUSING

The construction of new permanent housing is part of the reconstruction phase and no longer deals with housing problems encountered during the various relief stages immediately following the earthquake. Although pressure to build new permanent housing units as quickly as possible will always exist, more time is available for the planning and design process during the reconstruction phase, which contains longer-term objectives, in contrast to the immediate postearthquake relief period. However, in order to maintain continuity of the housing topic, the subject of new permanent housing will be covered in this part of the housing section rather than including it in the reconstruction commentary (the next major section in this chapter).

Unquestionably, architects and planners have important roles in this aspect of housing reconstruction goals. It is crucial to the entire process that planning and design professionals are prepared to work in this critical topic area after a major disaster. Many opportunities will occur at that time for their paarticipation in the planning and design of permanent housing units.

Tangshan, China. For sheer size, one of the most dramatic examples of postearthquake construction of new housing is found in the reconstruction of the city of Tangshan. In 1949 the city's population was about 470,000, but by 1976 it had increased to 1.6 million. In the 1976 Tangshan earthquake, over 95 percent of all buildings in the city, including housing units, either collapsed or were so severely damaged that they had to be abandoned.

One of the significant decisions made as part of the Tangshan reconstruction program after the earthquake was to reduce the population targets to a more manageable 1.0 million residents, a reduction of about 40 percent of the population. A second important decision was to make the provision of adequate housing one of the main goals of Tangshan's reconstruction. While the population in the damaged, older area of the city was held to 250,000, the design of a new residential district, planned to the north, was consciously planned on a cohesive neighborhood basis. It was divided into 118 distinct and small living communities, each of which could accommodate 5000 to 10,000 inhabitants. Support facilities, such as schools, a nursery, theaters, and shopping centers were provided in each neighborhood area. In each such neighborhood, rows of new rectangular blocks of mid-rise apartment houses (see Figure 9-11) were carefully spaced, with open areas between units, and interspaced with neighborhood centers and schools. The new apartment blocks are considered impersonal by some, and are undergoing personalization in the uses of the exterior balconies and open spaces near the buildings. Again, because of the culturally strong emotional attachment to land, gardens and private storage units are also appearing adjacent to the mid-rise buildings.

Mexico City. A second notable case history of new housing reconstruction in a large urban center is offered by the experience in Mexico City after the 1985 earthquake. The building sector most affected in terms of the number of structures damaged or lost was housing; about 68 percent of the buildings in this category totally

Figure 9-11 *New permanent multistory housing units in Tangshan reconstruction, 1976 Tangshan earthquake.*

or partially collapsed. The most dramatic losses in the housing stock were those in high-rise structures. (See Figure 9-12.) However, large numbers of housing units lost were in the central part of the city, albeit with lesser loss of life since these were generally low-rise buildings, many of them old, which had been converted long ago from single-family to multiple-family dwellings.

About 30,000 dwellings were destroyed and another 70,000 were damaged in the center of Mexico City. Residents were moved out of the damaged buildings during a period that averaged eight months. Twenty thousand provisional dwellings were built at this time and an additional 19,900 families were subsidized with rental assistance in other units. During the peak new housing construction period, over 100,000 people were living elsewhere. In the Tlatelolco housing development, which had 100,000 residents, a special program was instituted for reconstruction and rehabilitation. Eight of the buildings that suffered critical damage or collapse were demolished, 32 units were retrofitted and strengthened, and 60 additional units were repaired. During this reconstruction program,

Figure 9-12 Severely damaged and partially collapsed reinforced concrete high-rise building, 1985 Mexico City earthquake.

10,000 persons had to be temporarily relocated according to Francesco Casada (1987).

The massive housing effort in Mexico City was accomplished under several separate programs and phased over a years's time. The various housing programs that were accomplished during this period are described in Table 9-4.

RECONSTRUCTION PHASE

The postearthquake reconstruction period, which has a vital role in the total urban recovery phase, involves a complex process encompassing a multitude of considerations and alternatives when viewed in an urban context. Containing so many enigmatic facets and requiring a multidisciplinary approach, it should never be approached from an indiscriminate basis, but rather requires careful study. Although urban planning and design has often been cited as having a potential role in earthquake hazards mitigation efforts, in the United States it has had a relatively short history in the reduction of seismic risk.

TABLE 9-4 Postearthquake Housing Reconstruction and Rehabilitation: Mexico City Programs Following 1985 Earthquake

Program	Objectives	Number of Housing Units
Phase I	Existing dwelling stock built for government housing agencies regrouped and reallocated to earthquake victims	16,332
Tlatelolco	Reconstruction and rehabilitation	9,492
Phase II	Rebuilding and upgrading of damaged dwellings not considered by expropriation program	12,670
Jalisco (in the southwest)	Rebuilding of housing and relocation to northwest Mexico	3,181
Nongovernmental (NGO) organizations	Reconstruction and upgrading of dwellings on expropriated lots	4,854
Housing Units Reconstruction Program (HRP)	New housing reconstruction program	48,800
	Total:	93,515

Source: Iglesias, J. (1989). Reprinted with permission.

Within the last 15 years, several earthquakes throughout the world have impacted major cities severely enough to affect the economic and political stability of the region. Under such circumstances, it has been necessary to support reconstruction programs with outside resources such as those provided by the World Bank, UNESCO, the International Monetary Fund (IMF), and other international funding units. Reconstruction of major metropolitan centers following a severely damaging earthquake has become a global problem tied to an international scale of economics and development. Data indicate that large, high-density cities located in regions of high seismic risk, of which there are many, may well be dependent on assistance from the outside for postearthquake reconstruction efforts. Reconstruction after the 1976 Tangshan earthquake, in which China refused all outside assistance, will probably be the last time that a country will be able to deal with the crisis by itself.

Urban Technological Considerations

Urban centers encompass almost every imaginable kind of human activity: the larger the city, the more diverse and specialized these activities. Many urban centers find themselves subject to intense pressures of population growth and sprawling physical developments. Most experience problems associated with inefficient land use development, resulting in urban sprawl and a breakdown of effective transportation systems. In critical cases these pressures have led to a mismatch between service needs for the ever-increasing numbers of residents and the services available to them. This may quickly lead to gaps between a city's financial resources and its ability to meet social responsibilities in terms of public health and safety. Under such pressures, it is not uncommon to find seismic safety goals at the bottom of the priority list, especially in areas where the earthquake return period is greater than 100 years.

Seismic problems experienced by a city are dependent on: (1) the region or area in which it is located; (2) the characteristics and condition of its physical infrastructure, including the existing building stock; and (3) the size and composition of its demographical makeup. Hence, there is no single urban problem per se, just as there is no single, simple solution to the seismic question. Yet, because of the opportunities, diversity of services, and activities available, metropolitan centers will continue to offer desirable and alternative life-styles.

URBAN TECHNOLOGY IN EARTHQUAKE HAZARDS REDUCTION

The idea of introducing seismic considerations in the reconstruction of a city following an earthquake can be traced back about three centuries. Even in 1667, substantial domestic resources and use of foreign balance of trade funds available for the postearthquake reconstruction of Dubrovnik, located on the Adriatic coast of Yugoslavia, converted the original town of ancient Roman influence into a dynamic and attractive baroque seaport that has survived many seismic events since then. According to Professor Stephen Tobriner (1982) after Lisbon experienced a severe earthquake in 1755, the marquis of Plombal introduced a reconstruction plan that adopted some urban planning elements to make the city safer during future seismic events. These elements included components such as height limitations of buildings, regular dimensions of blocks for easier access and egress, recommendations for width of streets, appropriate distances between buildings, and others.

Following the Lisbon experience, architectural and planning antiseismic measures were adopted in the city of Noto, Sicily, during the reconstruction phase that followed an earthquake in the late eighteenth century. For a long period since then, few steps had been taken to formalize earthquake safety elements in the reconstruction planning and design of urban centers. More recently in China, the reconstruction plan for the city of Tangshan, which had to be totally rebuilt after the 1976 earthquake, successfully reintroduced urban planning considerations to reduce future seismic hazards. The USSR is following a similar direction for the reconstruction of urban centers destroyed during the 1988 Armenia earthquake.

Major cities still remain extremely vulnerable to earthquakes because much of its building stock may have been constructed before the knowledge and science of earthquake engineering had become fully developed. Even in the United States, as indicated in Chapter 7, building codes are not retroactive despite the fact that major code changes are made to seismic provisions after a severe, damaging earthquake. Several were made after the earthquakes in 1964 in Alaska and 1971 in San Fernando. Even the 1985 Mexico earthquake was responsible for the addition of a fourth soil factor, S4, in the site coefficients for the seismic provisions in the 1988 edition of the Uniform Building Code. The implication is that if a major code change is introduced in a particular year, such as new ANSI or NEHRP provisions, the majority of buildings, technically speaking, may not meet the new standards. Consequently, as indicated previously, at any given time the vulnerability of metropolitan centers to seismic events may remain a significant problem.

A large-scale approach to the problem is necessary, especially since urban technology involves the preparation of general plans for future growth and change in urban areas. This includes consideration of open spaces, land use planning, circulation systems, access/egress routes, building topology, and implementation of a general plan. The least that can be done is to introduce a seismic safety element into the general plan in recognition of the earthquake hazards existing in the area. Implementation of comprehensive urban planning procedures may be used to influence the reconstruction process and growth of existing cities and thereby reduce their vulnerability to future earthquakes.

Reconstruction Options and Strategies

A course being recommended by leading Western planners is that a city's social and economic structure first be recaptured and restored before following up with large-scale physical planning and rebuilding activities. The postearthquake reconstruction phase belongs to the latter rebuilding stage. In this rebuilding stage, it is recognized that there are three basic options for the reconstruction of metropolitan centers severely damaged or destroyed by a major earthquake (see also Figure 9-13):

1. Rebuilding on the same site, no change in location,
2. Decentralizing services and decreasing high-density congested areas through the construction of new satellite districts, and
3. Abandoning the old site and moving to a new location.

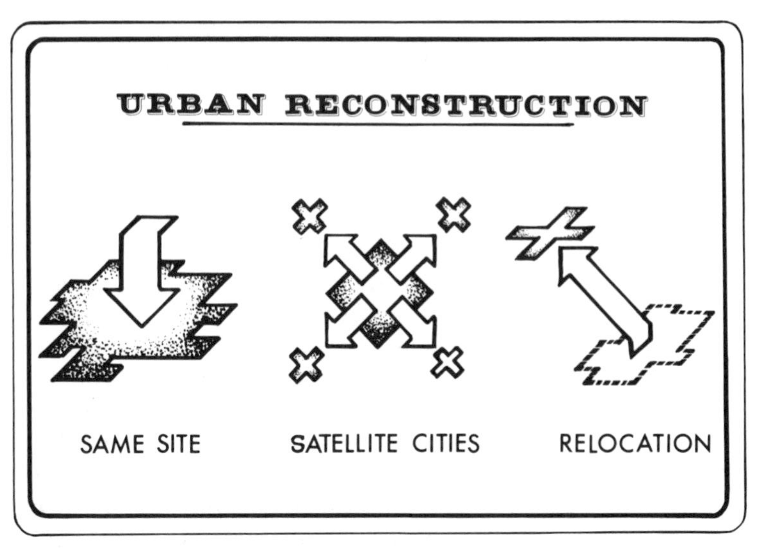

Figure 9-13 Three postearthquake urban reconstruction options. *Source:* Irene R. Lagorio.

Normally, active social and economic pressures after an earthquake will influence reconstruction along patterns that already existed before the event. There are exceptions of course, such as those found in societies with strong central government systems that dictate policy. However, in general, reconstruction programs that recommend relocation options for specific sections of the community quickly raise concerns of social equity. Since existing patterns of the urban fabric may be visibly severed during the process, total and final recovery can be impeded. How and at what scale the reconstruction process proceeds is directly dependent on the social system, cultural traits, political context, and economic structure of the affected society itself. But for the purposes of this book, in the actual case study examples presented later emphasis will be deliberately placed on the urban planning and design aspects of postearthquake recovery.

OPTION NO. 1: CITY OF SKOPJE, YUGOSLAVIA, 1963 SKOPJE EARTHQUAKE

Not counting other notable historic examples of urban postearthquake reconstruction during the eighteenth century, one of the most dramatic efforts in rebuilding a city can be found in the case history of Skopje, which was destroyed by a catastrophic seismic event in 1963. The city then had a population 200,000. The devastating earthquake, which registered a 6.1 magnitude on the Richter scale, killed 1070 people, left 150,000 homeless, and demolished or severely damaged up to 75 percent of the buildings. Adobe structures in the center of the city were unable to withstand even the moderate ground shaking that occurred.

Soon after the first month passed, central government officials considered all options in the planning of a reconstruction program. Option 3, complete relocation of the city to a new site, was seriously considered. According to M. R. Greene (1987), however, after many studies on soil conditions and the seismic vulnerability of the region were concluded, it was finally decided that no alternative site existed in the Macedonia republic that would be more appropriate for the capital city, and that all other alternative sites under consideration were equally vulnerable. A conscious decision was therefore made to rebuild on the same site as long as the reconstruction program included steps to mitigate seismic hazards. One of the motivating factors behind the decision was that in economic terms 80 percent of Skopje's financial capital (in transportation facilities, public utility services, operational factories, repairable building stock, and undamaged structures) was still in place.

Systematic soil borings were taken and seismograph records of the many aftershocks were made in order to determine areas in which the city could be appropriately and safely extended if need be. A master plan, prepared by an international group of architects and planners, was approved in 1965, two years after the earthquake. The massive international stream of assistance into Skopje included help from the UN, which, in coordination with the central government, introduced a UN Urban Project as further support to the reconstruction program.

It was also decided to give the highest priority to the replacement of the housing stock. Fourteen thousand new permanent dwelling units were completed in the year following the earthquake. More than 16,000 apartment units were repaired and reoccupied soon after the earthquake. Redevelopment and reconstruction of the new City Center complex were postponed until additional land could be purchased to assure that Skopje would be an impressive capital with new planning and design patterns worthy of note.

Another interesting feature of the Skopje reconstruction effort was the decision to hold an international design competition under the sponsorship of the UN and Yugoslavia's central government. This was the first time that such a competition had been held for a postearthquake recovery project. The main purpose of the design competition was to obtain alternative ideas that could be used in the finalization of the reconstruction plans for the new City Center complex.

The competition was an invitational one, in which eight design teams, four Yugoslavian and four from other countries, were invited to present their recommendations and designs. The two best designs selected by the international jury were one by architect Kenzo Tange of Tokyo and the other by the Croatian Institute of

Town Planning in Zagreb, with the institute having final responsibility for the final design and its implementation. In the end, it was apparent that the Kenzo Tange plan had heavily influenced the direction taken by the final plan for the new City Center.

Although Kenzo Tange's plan was awarded a first prize, a slight controversy ensued that claimed that his solution was impractical, unrealistic, and out of proportion to the limited resources available to Yugoslavia at the time, despite economic assistance from the UN. There were comments that the design was too large in scale from that which was warranted. However, other comments from the jury that evaluated all the projects indicated that Tange's design was brilliant in its solution of transportation linkages and social communications systems. In particular, positive mention was made of the excellent relationship between the Transportation Center and the main entrance to Skopje, provision of many open spaces, and the overall decentralized concept, which effectively mitigated some of the deficiencies of the congested, more vulnerable parts of the old city.

Twenty-six years later, owing to the rapid growth of the city, residential construction proved to be the largest part of the reconstruction effort. The plan for the City Center, especially the city gate and walls inspired by the Kenzo Tange design, has been scaled down and only partially implemented in terms of being faithful to the spirit of his plan. The second and third construction stages of the general urban plan have yet to be finished.

The Transportation Center, which was envisioned as a complex containing a railroad station, bus depot, and airport terminal, and was intended to serve as a major social focal point and transportation hub, only incorporated a few of the intended functions. Nevertheless, the reconstruction of Skopje still represents a bold approach and was successful in many ways. It is especially noteworthy for its desirable residential units, wide streets, and the implementation of several other basic seismic safety features in its original concept.

OPTION NO. 2: CITY OF TANGSHAN, CHINA, 1976 TANGSHAN EARTHQUAKE

Tangshan is one of the major industrial cities in the Hopeh Province on the main railroad line that connects Beijing, Tianjin, and the coast. It was founded in early 1870 when the underground coal deposits in the area assumed great importance. The coal mines of Tangshan developed into a significant industry and became a major economic resource and center of employment. Unfortunately, over the years, as its industrial capacity increased over 220-fold, further development of the city was not controlled. Tangshan became a high-density, congested urban area with complex circulation patterns and narrow streets. There was only one main route that led into the city and only one way out.

The 1976 earthquake that destroyed and severely damaged over 95 percent of the buildings left the city in ruins. There were 250,000 deaths and more than 80,000 injuries that required hospitalization. (See Figure 9-14.) Potable water service was not resumed until 12 days after the earthquake. The heavy damage provided an opportunity to start over, from scratch, so that all three of the reconstruction options were feasible.

Because of the economic and industrial importance of the coal mines, it was finally decided that Tangshan should be reconstructed in the same general area, and consequently, relocation to a new location was rejected. Yet it was recognized that building in the same location could potentially duplicate the same pre-earthquake problems: high-density, congested neighborhoods; poor circulation; building over major fault traces; construction on poor soils areas subject to surface rupture; and general air pollution from heavy industrial concentrations. In addition, as China is uninhibited by Western policies of private ownership of land and property, the

Figure 9-14 Severely damaged and partially collapsed low-rise masonry housing units, 1976 Tangshan earthquake. *Source:* Ministry of Construction, People's Republic of China.

central government had relatively free reign to depart from the old pattern of the city in order to plan for a radically new layout and form. Thus, rather than totally rebuild on the old site and duplicate existing land use patterns, Option 2, decentralization of services and construction of new urban satellite communities, was selected for the reconstruction plan.

Tangshan's reconstruction plan called for three interdependent areas approximately 15 miles from each other: (1) the redesigned site of the old original city; (2) a new satellite industrial district to the east developed and regrouped around the coal mines, and (3) a completely new satellite district to the north developed as the principal residential district, called Fengren. (See Figure 9-15.) The original reconstruction program limited the total population of the new city, with its three sections, to 1 million, although that target was already surpassed several years later.

The section where the old city stood is a predominantly large open area with parks, recreation areas, and additional landscaped areas between buildings. It contains the commercial and cultural core with few and adequately spaced department stores, a 16-story tourist hotel, medical facilities, a few old damaged buildings to serve as memorials to the earthquake, and a large formal park. The population is limited to about 250,000.

The industrial part to the east around the coal mine areas was purposely kept separate from the other two parts because of environmental concerns about pollution. It contains all the new factory and heavy industry facilities (the old ones having been destroyed by the earthquake) and additional housing for the workers.

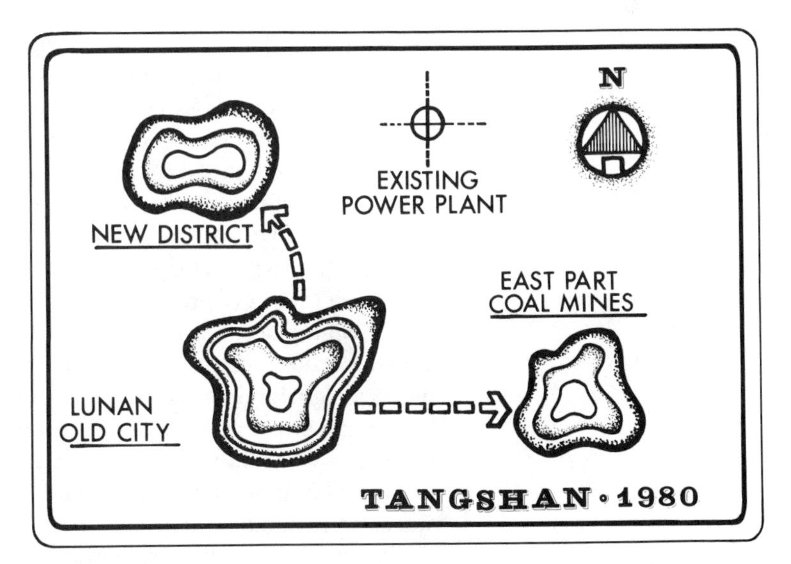

Figure 9-15 Diagram indicating postearthquake reconstruction concept of new Tangshan following 1976 earthquake. *Source:* Irene R. Lagorio.

The new principal housing area to the north, Fengren, also contains neighborhood shops, schools, theaters, preschool nurseries, and open spaces in each residential quarter. Plans were to keep the population in this section to approximately 500,000 to 600,00 residents, but by 1985 it had already increased to about 750,000.

Many seismic safety features were incorporated in Tangshan's urban plan. The transportation system was meant to allow for easy access and egress in all directions (the old system was tortuous, with narrow confined streets and alleys, and only had a couple of main entrances and exits from the congested city). Consequently, two new main arteries about 200 ft wide were developed as the main circulation system. Within each right-of-way, 50 ft are used for autos and buses, 5 ft serve as a divider strip, 25 ft are allocated to bicycles, and 20 ft are devoted to pedestrian traffic. All new major buildings are along these two main arteries. The old railroad line has been rerouted to avoid the prominent existing earthquake fault traces and mines of the coal-bearing grounds.

The new urban form has been based insofar as possible on geological investigations, identification of potential seismic activities, microzonation techniques, and urban vulnerability studies. Major buildings and important service facilities are sited on shallow bedrock of high bearing capacity. Ample space is used between buildings to facilitate evacuation and mitigate the potential of casualties from future earthquakes. Wide streets will prevent traffic congestion and provide adequate and clear (debris-free) access to future potential earthquake-damaged areas.

OPTION NO. 3: CITY OF VALDEZ, ALASKA, 1964 GREAT ALASKA EARTHQUAKE

The city of Valdez, Alaska, was originally located at the end of a immense glacial outwash. During the 1964 Alaska earthquake, which registered a 8.4 magnitude on the Richter scale, 100 million cubic yards of soil moved, submarine landslides occurred, a sea wave inundated the city, and ground fissures opened up everywhere. Its waterfront pier was literally wiped out and structures were heavily damaged.

After the earthquake, it was obvious in no uncertain terms that Valdez sat on a very precarious location. If it was decided to rebuild in the same place, Valdez would not have been eligible for federal aid, since the location did not meet federal safety guidelines. No mortgage guarantees and no loans would be available. In effect, the federal government indicated that it would give support funds only if the city were moved to a safer site; otherwise, the residents and local government of Valdez would have to pay for all the reconstruction programs themselves.

Under these conditions, Option 3, the most drastic of all possible reconstruction programs, was the only viable alternative. Fortunately, Valdez was a small city with a very small population at the time. It was moved to a totally new, more stable location immediately to the north of the old site. A complete, new city was designed and built in a formal gridiron pattern on the new location, which was situated at a higher elevation to minimize the hazard from future submarine landslides and sea waves. The net cost to the federal government was $330 million. To this day, it is still the only example of a complete relocation of an urban area after an earthquake.

One word of caution is necessary when evaluating the impact of the relocation and reconstruction of Valdez. At the time of the earthquake in 1964, the population of Valdez was only slightly over 1000. In contrast to the monumental effort in the reconstruction of Tangshan, where over 1 million persons were relocated to the new satellite districts, the relocation of Valdez and its new layout plan for its small population were relatively easy. Of course, on the other hand, the reconstruction of Tangshan was made simpler by the central government format and decision-making process found in China. Even today, the new Valdez is still a small city, having a population of about 3000 in 1985.

POSTEARTHQUAKE RECONSTRUCTION IN ARMENIA, 1988–1989

The 1988 Armenia earthquake, which measured 6.9 on the Richter scale, struck on December 7, 1988. According to seismologists, the earthquake was caused by the tectonic system where the Eurasian and Arabian plates converge in the Caucasus region near Armenia. Within the area, evidence indicates that clay soils amplified ground motions—the same problem that occurred in Mexico City. The natural period of the precast reinforced concrete buildings, which were 5 to 17 stories high, coincided with the natural period of the soils, leading in many cases to structural failure. The problem was compounded by poor on-site quality control of construction details and materials. (See Figure 8-1.)

The structures damaged beyond repair by the earthquake were the typical precast reinforced concrete building systems built in the USSR during the 1940s to 1970s. Some were as tall as 16 or 17 stories. The epicenter was almost directly located beneath the small city of Spitak, which probably accounted for the city, with a population of about 13,000, being completely destroyed. A larger city, Leninakan, with a population of about 250,000, was also hit hard, with up to 75 percent of all residential units destroyed or severely damaged.

Spitak is being completely rebuilt and relocated 5 miles to the southwest of its original site, which is being abandoned (Reconstruction Option 3). The decision to relocate the city was made on a hazards mitigation basis, because the original site was crisscrossed by active fault zones to which the epicenter of the 1988 earthquake was attributed. The new town plan of Spitak incorporates all the appropriate amenities: new permanent housing, schools, recreation areas, a commercial area, a central marketplace, new transportation systems, and local government facilities. To avoid the recurrence of resonance between the natural period of the structures and soils during future earthquakes, new buildings are being limited to four stories, with a maximum of five stories in a few special situations.

The city of Leninakan is being rebuilt in three stages and decentralized at the same time. In undertaking the reconstruction program, the residents first sought assurances that all seismic problems had been solved before giving approval to rebuilding. Under Stage 1, new permanent housing units for 150,000 residents are being constructed in a new area on the northwest side of the old city. Stage 2 will consist of the restoration of the old historic district, and Stage 3 will entail reconstruction of the commercial area at the northern part of the city. New buildings are designed as poured-in-place concrete structures with stiff shear walls, all quite different from the precast concrete buildings that collapsed during the earthquake. (See Figure 9-16.)

Reconstruction costs for all damaged areas in Armenia total about 10 billion Russian rubles. It is interesting to note that this figure is said to be on the same level as the reconstruction costs of

Figure 9-16 Architectural model indicating postearthquake reconstruction plan for Leninakan, USSR. *Source:* Arnold (1989). Reprinted with permission.

the Chernobyl nuclear disaster. This sum amounts to about $1.5 billion in U.S. currency.

Other Basic Considerations in Reconstruction Planning

Another consideration is to determine which urban technology procedures can be effectively and appropriately utilized during the reconstruction phase for the reduction of future earthquake hazards. As a subelement of potential reconstruction policy, one question is what is to be done with damaged buildings that are still standing but no longer suitable for occupancy in their damaged state in metropolitan centers.

Using the 1985 Mexico earthquake as a base, policy on the general treatment of damaged buildings as a generic problem is worthy of review. A set of multistory buildings in Mexico City, assessed over a three-year span following the 1985 earthquake, are presented for potential approaches in answer to this problem. For practical purposes, the commentary is limited to the assessment of recent buildings; its intention is not to deal with older, historic monuments, which represent a special problem involving historic preservation and conservation, another important topic having a significant role in reconstruction following earthquake.

As indicated in several publications, after the 1985 earthquake, Mexico City became a natural full-scale "seismic laboratory" for the testing of existing buildings of all types and ages subjected to long-period ground motions. The mix of buildings in Mexico City reveals some interesting data. In 1852 the Spanish started to build the colonial city with unreinforced stone masonry structures of one, two, or three floors. Many churches were to be found in the city at that time, including the older part of the cathedral with its two towers about 52 meters high, located in the Zocalo. During the 1930s, buildings of modernistic construction, up to 17 stories high, began to appear. They were mainly designed by static analysis and had little or no flexibility incorporated in their design. In 1956, the tallest building in Mexico City, the 44-story Latinoamericana Tower (which suffered little damage during the 1985 earthquake), was completed; it was built with a pile foundation system that extended all the way through the poor lakebed soils into firm ground. Many other multistory reinforced concrete and steel-frame buildings of different sizes and construction had been completed since then to produce the Mexico City seen before the 1985 earthquake.

From the overall urban perspective, these diverse buildings had different seismic performance characteristics in reaction to lateral load input motions induced by earthquakes. Damage patterns resulting from the severe ground shaking of long duration in Mexico City's complex urban environments related directly to the number and types of buildings at risk as well as to their location relative to the lakebed soils. Approximately 5700 buildings located in the central city were damaged, many severely, or destroyed. Of the total,

485 suffered total or partial collapse, and 271 suffered major structural damage. Another set of statistics indicate that a total of 950 buildings were demolished and 2300 were severely impaired. Damage to 2450 buildings was listed as moderate to minor. Less damage was experienced by buildings in the transition zone and hill districts, versus those directly in the lakebed area.

As a result of the 1985 earthquake, a new building code was introduced. Established in 1987, it mandates that all previously existing structures in Group A (very important structures, the failure of which could lead to major casualties, large financial impacts, and/or cultural losses) that do not meet the 1985 or the 1987 building code standards be strengthened so as to comply with the new requirements. This new stipulation is also aimed at existing structures in Group B (ordinary structures, which were seriously damaged in 1985) for which there are doubts concerning their safety with respect to failure or serviceability.

NEW ZONING REQUIREMENTS

Additionally, new provisions for off-street parking were introduced as part of the emergency ordinances legislated. One parking space is now required for each 30 square meters of floor office space. For apartment buildings, 1 to 3.5 parking spaces per apartment floor space ranging from 60 sq m to 250 sq m are required. Also as part of new zoning restrictions, new general building set-back requirements have been established for certain streets and areas. Typically, these new ordinances state that the height of new buildings cannot exceed the width of the street, and that an open space of at least 15 percent of the building height must be set aside in the rear of the property. In addition, from 20 to 30 percent of the property must be developed as open space. A special ordinance for the historic center of Mexico City limits the height of any new building to four stories. It is interesting to note that there were many 5- to 10-story existing buildings, including the 44-story Latinoamericana Tower, located in this area before the earthquake.

DAMAGED BUILDING REHABILITATION OPTIONS IN THE RECONSTRUCTION PHASE

In the rehabilitation of damaged buildings in Mexico City after the 1985 earthquake, at least five general approaches were used during the reconstruction stage:

Option	Rehabilitation Strategy
1	Repair/strengthen
2	Reduce building mass
3	Demolish/rebuild anew
4	Demolish/create open space
5	Preserve damage status as an earthquake memorial

Options 1 through 4 have been used extensively in the rehabilitation of damaged buildings. It is interesting to observe that while Option 5 was visibly used in the reconstruction plan of Tangshan, China, it was not identified as an immediate choice in Mexico City, where such preservation is occurring on a smaller scale. Data collected are still being analyzed to determine the frequency of use of each of the five options in Mexico City for correlation with the total number of damaged buildings.

BUILDING REPAIR/STRENGTHENING METHODS

A field sampling of damaged buildings was reviewed and assessed as part of an effort to identify methods used. Several methods were examined that served as generic procedures in the repair and strengthening of damaged buildings. The most common methods used are shown below:

Method	Description
1	Steel-jacketed columns
2	Addition of shear walls
3	Addition of diagonal bracing
4	Stiffening of horizontal floor Diaphragms
5	Reinforcing foundations
6	Epoxy in minor reinforced-concrete-frame cracks

It is not unusual that two or three of these methods be used in the rehabilitation of a building. As fewer steel-frame buildings existed in Mexico City, most of the case studies analyzed were engineered medium-rise reinforced-concrete-frame buildings. As methods for the repair and strengthening of unreinforced masonry structures were previously covered in Chapter 7, low-rise historic masonry buildings are not presented in this section.

As previously indicated in Chapter 8, over 42 percent of the heavily damaged buildings in Mexico City were multistory corner buildings highly susceptible to unforeseen torsional effects caused by the earthquake. Of this group, flat-iron buildings (located at acute-angle street intersections) in particular exhibited damage due to torsional effects. Typically, when possible, Method 4 (stiffening of horizontal floor diaphragms) was used to strengthen the floors of these buildings to minimize the damaging torsional effects of future earthquakes. (See Figure 9-17.)

Approximately 40 percent of the damaged buildings were left unoccupied for about two years while an appropriate rehabilitation method was determined and reconstruction financing arranged. In December 1988, in the central historic district of the city, three years after the earthquake, severely damaged reinforced concrete-frame buildings with temporary bracing systems still remained standing but also abandoned, with their facades being stripped

Figure 9-17 Postearthquake repair and strengthening of reinforced concrete flat-iron building, 1985 Mexico City earthquake.

away while their ultimate fate was being decided. (See Figure 9-18.) Such extended waiting periods represent a sizable loss of resources, resulting in a significant economic impact over the years.

DEMOLITION OF DAMAGED BUILDINGS

Through government powers related to public health and safety, expropriation, and/or eminent domain, severely damaged buildings and those that suffered dramatic collapse beyond the realm of rehabilitation were condemned and demolished, and all debris was removed from the sites. Data on the exact numbers of buildings in this category are still being collected and analyzed.

Some sites are still empty, and having been fenced in, remain unused. Others, distributed throughout the city, have been converted to open space, with many used as "miniparks." For example, the site of the collapsed Regis Hotel has been used effectively as an extension to the adjoining outdoor plaza and central park, the Alameda. A small new building on the site has been constructed to serve both as an earthquake memorial and public museum; it houses the Diego Rivera murals rescued from the severely damaged Hotel del Prado across the street, which was demolished in 1988.

The three critically damaged 21-story steel-frame towers (one of which overturned during the earthquake) at the Medical Service Center Complex above the Pino Suarez Subway Station have been demolished. As heavy ballast to hold down the subway station below, the old center has been replaced by a new massive two-story low-rise reinforced concrete complex with an outdoor terraced plaza and community services building surrounding an additional open space in the area. (See Figures 9-19 and 20.)

OTHER BUILDING REHABILITATION OPTIONS AND METHODS

Building rehabilitation options and methods used to improve a building's seismic performance, other than those described above,

Figure 9-18 Empty damaged building waiting repair or demolition, 1985 Mexico City earthquake.

Figure 9-19 Postearthquake remains of three high-rise steel-frame buildings above Pino Suarez Subway Station, 1985 Mexico City earthquake.

were observed. Option 2 (reduce building mass) and Method 5 (reinforcing foundations) were used effectively.

A major office/commercial sales building located on the Paseo della Reforma near the Cristobol Colon Plaza represents an excellent example of reducing the mass of a building in order to improve its seismic performance. Originally an 11-story building, the top

Figure 9-20 New reinforced concrete community center and terrace park above Pino Suarez Subway Station, Mexico City, 1988.

Figure 9-21 Removal of upper floors to reduce mass of multistory building in Mexico City following the 1985 earthquake.

five stories were systematically removed floor by floor. (See Figure 9-21.) Currently it is completely occupied and operational. The same method was used in several of the residential buildings at the Tlatelolco complex.

Rehabilitation Method 5 (reinforcing foundations) was also used extensively in the Tlatelolco complex, primarily on the large-scale residential building type, one unit of which had overturned during the earthquake. At one unit, massive heavily reinforced concrete foundations were added as stabilizing arms beyond the building's outside perimeter to avoid overturning of the superstructure above. (See Figure 9-22.) Where minor damage occurred in smaller reinforced concrete-frame buildings, high-pressure epoxy was injected into all visible cracks to make repairs (see Figure 9-23). Addition of new reinforced concrete shear walls to strengthen reinforced concrete-frame buildings, Method 2, was used extensively (see Figure 9-24).

BUILDING REHABILITATION COSTS AND ECONOMIC IMPACTS

In Mexico City, significant amounts of funds are being spent on the rehabilitation of the existing building stock damaged by the earthquake. The entire rehabilitation effort is a complex process, owing to the large number of structures requiring attention and the fact that each building represents a unique problem. At the moment, common solutions adaptable to all buildings do not exist; each structure has to be addressed on a case-by-case basis. As every seismic rehabilitation case embodies a unique combination of preconditions and special procedures, cost data are not relatively applicable.

Figure 9-22 Strengthening multistory housing unit in Mexico City Tlatelolco area by extending reinforced concrete foundations.

Figure 9-23 Epoxy repair of damaged reinforced concrete column.

At present, complete reconstruction statistics on cost relationships and financial data bases do not exist. However, as a partial example of the magnitude of public funds initially expended to date for the reconstruction program in Mexico City, a total of $573 million of World Bank funds have been used. This amount consists of $400 million of new funds plus a $173 million reallocation of funds from previous World Bank projects in Mexico City. These funds include expenses for demolition, debris removal, relocation of management agencies, retrofit of public school buildings, repair and reorganization of hospital buildings and services, and decentralization of medical system buildings.

Total private cost data on the seismic rehabilitation of privately owned damaged buildings in Mexico City are still being collected in aggregate form for reporting on a building-class basis. In certain cases, it has been noted that cost data are skewed to the high side as building owners are openly seeking general performance guarantees as hedges against further damage by future earthquakes. In addition, because new zoning ordinances, property development restrictions, and higher building performance standards have been instituted since the 1985 earthquake, owners are willing to spend more funds than typically required for rehabilitation rather than satisfy the new building code requirements instituted in 1987 (described earlier).

Figure 9-24 Strengthening damaged building in Mexico City by adding new reinforced concrete shear walls.

SUMMARY

The rehabilitation of earthquake-damaged existing building stock is an important part of the reconstruction process and is a means of preserving the urban fabric. As site relocation solutions (as for the small city of Valdez, Alaska) are not appropriate for most urban centers, building rehabilitation needs warrant careful attention before damaged structures are immediately demolished and lost forever (as was done in Coalinga).

Additional economic data needs to be collected, assembled, and collated as a data base to yield a sound understanding of the financial impact of building rehabilitation on the fiscal resources available for reconstruction. In Mexico City the final, total economic impact on both public and private sectors has yet to be clearly and comprehensively totaled, so the information cannot yet be applied to any other reconstruction efforts following a major earthquake. The reticence of governments, private corporations, and insurance companies to release total and final reconstruction costs is quite common after a major disaster.

CHAPTER

10 EARTHQUAKE HAZARDS MITIGATION PROCESS

GENERAL REVIEW OF THE MITIGATION PROCESS

Architects must be prepared to take advantage of earthquake hazards mitigation programs as an extension of their professional services or find themselves falling behind other representatives of the design professions. Earthquake hazards mitigation involves a distinct process, the goals of which are to reduce the potential vulnerability exposure of lives, property, and resources to the effects of earthquakes by taking preventive actions. The primary emphasis of an effective mitigation process should be on actions taken before the earthquake rather than after. However, at the outset it should be apparent to all that owing to the diversity, complexity, and variable recurrence periods of earthquakes, as well as the public's perceptions of exposure to seismic hazards, hazards mitigation is difficult.

Earthquake hazards mitigation infers the taking of intentional actions to modify, reduce, or prepare for the risk of a potential disastrous seismic event of an uncertain recurrence interval. These actions, personal or impersonal, may be incorporated into four levels of intervention:

1. Halt the earthquake itself from occurring,
2. Avoid areas of high seismicity or move to a new location that is not seismically active,
3. Predict the occurrence of an earthquake in advance, and
4. Prepare for and/or reduce the impact of our vulnerability to earthquakes.

Plainly, the first mitigation action, though very intriguing, is impossible to achieve at this point. At least for the moment, the possibility of impeding an earthquake from occurring through intervention by mortals appears unrealistic. The second may not be totally viable economically nor environmentally, since the majority of habitable regions in the world are in seismically active zones. Furthermore, many people would refuse to move, as enforced relocation would be interpreted as a violation of human rights and politically unacceptable. This leaves the third and fourth actions as the most promising mitigation efforts.

On an international basis, the difficulty of achieving effective earthquake hazards mitigation programs is due to the fact that various countries throughout the world have problems, whether cultural, economical, or environmental, that are unique to them. This makes the transfer of technical information difficult for preparedness and response-planning activities. However, in all cases the prime goal of mitigation efforts is to improve community public health and safety, whether it be on a local government, national, or international basis.

An integrated approach involving multidisciplinary inputs is necessary. Efforts by the public and public officials, several professional disciplines, economic analysts, and social scientists are required to make the mitigation process successful. Although it is acknowledged that mitigation requires contributions from all groups, the elected or appointed public official has perhaps the most significant role. Simply stated, it takes a proficient legislative approach and/or cooperation of public officials for major mitigation programs to be persuasive and accepted.

ASSESSMENT OF SEISMIC RISK

Before undertaking a mitigation effort, an assessment of the risk involved is required as the first step in the process. Any mitigation program placed into effect before an assessment of the risk is made is of no value, as it would lack a thorough understanding of the problem.

First, above all, the earthquake hazard must be identified and put into perspective relative to the exposure and vulnerability of the study area. As the process of identifying the seismic risk of an area depends on a number of variables, base input is needed from various sources at this point: the seismological, geotechnical, engineering, architectural, urban planning, socioeconomics, and demographical sectors. An overall assessment of the total risk can only be achieved by a balanced evaluation from all the disciplines involved.

A similar approach to that used in the case of reducing seismic risk for a single building, explained in previous chapters, can be taken to address the problem of risk reduction on a city and regional scale. In terms of regional planning relative to seismic risk analysis, risk is also defined as a combination of four principal factors: (1) hazard, (2) exposure, (3) vulnerability, and (4) location. As before, the hazard includes all potential geological hazards in the region that influence the earthquake's potential impact: ground rupture, ground failure, ground shaking, liquefaction, landslide, tsunami, and other collateral effects such as fire following earthquake, dam failure, and toxic chemical release.

Exposure refers to the degree or level of being subject to the hazard on a regional basis, including the anticipated impacts on public health and safety in face of the hazard. Again, regional vulnerability is perceived as a measure of the contingent damage or environmental performance of the study area on a large scale relative to its exposure. And, finally, location relates to the general position of the total regional study area to a potential earthquake source. As before, all four of these factors are evaluated and factored on a territorial basis to determine the level of seismic risk faced by the total study area under assessment. An area ranked as having a high seismic risk would have prohibitive degrees of hazard, exposure, vulnerability, and location scores. On the other hand, a potentially vulnerable area may not be at high risk if its location relative to an earthquake source is benign, and its regional exposure to being subject to a hazard is minimal.

By taking all these data into account, it is obvious that earthquake hazards mitigation programs should not be developed or implemented on a large-scale regional basis until a comprehensive seismic risk assessment is completed. To do otherwise could lead to some embarrassing situations and misuse of public funds if the risk analysis is flawed.

Voluntary and Involuntary Risk

Risk may also be classified from a personal perspective as to whether it is assumed on a voluntary or involuntary basis. For example, with all the information made available to the public by casualty insurance companies on the risks involved in driving an automobile, the risk is classified as voluntarily assumed. In risk assessment of earthquake hazards, Chapter 8 indicated that single-family detached residences are exempt from the Special Studies Zones Act in California, the land use ordinance that controls construction in fault-line zoning areas and identifies the zone as hazardous in case of an earthquake. Accordingly, in that case, if a couple consciously decides to file for a variance, and builds and occupies a single-family dwelling on property within a special studies zones area, they would be assuming a voluntary risk in being personally responsible for their own life safety and that of the entire family.

The legal aspects of this can be very significant, for if at a later date another person unknowingly purchases that residence without receiving a written disclosure from the seller that the dwelling is located in an earthquake-hazardous zone, the new buyer would have been placed in the situation of assuming an involuntary risk. After a severe earthquake, if any casualties or property loss occurred in or to the dwelling, the new buyer, having assumed the risk involuntarily by not having been clearly informed of the hazard, could have a right to seek legal retribution. Technically, the involuntarily assumed risk would be similar to an employee unknowingly working in a building known to the employer to be an earthquake-hazardous structure. In this day and age of liability problems, it is essential that architects be aware of the ramifications of building within well-defined earthquake-hazardous zones or being engaged in a project limited to interior design work for an earthquake-hazardous building owned by the client. Things can become quite sticky in this regard after an earthquake that results in extensive property damage, injuries, and life loss.

Risk Management Relationship to Hazards Mitigation

A general relationship exists between hazards mitigation goals and risk management in the broad sense. When any administrative director of a community services office examines potential losses from a possible disaster facing a jurisdiction and makes decisions to preserve its assets, or reduce the threat, that individual is operating within the duties of a risk manager.

The power of hazards mitigation efforts rests with the fact that if, because of a disaster, public money is diverted to nonproductive uses, such as the replacement of damaged property or payment of excessive liability claims, that money is no longer available to provide fully for services perceived to be necessary by the community. Consequently, on a long-term basis, if earthquake hazards mitigation efforts are successful in reducing damage levels, casualties, and property loss, the resulting savings can be put to a more productive use in meeting societal demands.

Contingency Planning

Contingency planning may also be classified as a form of earthquake hazards mitigation and risk management. It represents an anticipatory process in which stock is taken of a potential emergency event, such as an earthquake, even though its occurrence is uncertain. It is intended for use in circumstances not completely foreseen.

One implication of contingency planning, however, is that it is dependent on the expectation of something happening. As a planning process, it represents a preevent tool that is held in abeyance until the event occurs and then it goes into action. Although its objectives may be the same in eventually diminishing property loss and reducing further casualties, it differs somewhat from preparedness planning, which goes into preparatory action even in the case when an event may or may not occur.

MITIGATION RESEARCH

The nucleus of hazards mitigation rests with research. The process involves a progression starting with a hypothesis, then going on to research to test the hypothesis, and finally implementation if the testing is successful. In the vast field of earthquake engineering research, in which more architects should be involved, the most significant element is experimental testing in the physical laboratory.

In the testing laboratory, anything can be evaluated under actual test conditions, not only tests of the seismic performance of basic structures and structural components, but also nonstructural elements, architectural configurations and layout, lifelines, base isolation units, energy dissipation systems, and even search and rescue methods as well. In the United States there are many organized earthquake engineering research units and experimental testing laboratories working on the testing of new ideas. (See Figures 10-1 and 10-2.) On the West Coast the Earthquake Engineering Research Center (EERC) at the University of California, Berkeley, was founded in 1945 and its Simulation Laboratory dedicated in 1962. In 1985, the National Science Foundation (NSF) funded a National Center for Earthquake Engineering (NCEER) on the East Coast at the State University of New York at Buffalo.

The most dramatic testing device located at these research units is typically the dynamic shaking table driven by computer programs run with digitized seismograph records of actual earthquakes simulating the ground motions. The records of the 1940 El Centro, California, earthquake are often used for this purpose because at the time of the earthquake, seismographs were able to obtain clear records of the ground shaking that ensued. The EERC Simulation Laboratory's large shaking table at Berkeley, for example, was used to test base isolation components for use in the design of highway bridges, and later experiments were conducted to test the performance of steel industrial storage racks for their performance under earthquake loading.

Japan has a large testing facility at the Building Research Institute in Tsukuba that can test full-scale buildings up to six stories high. Rather than a large-scale shaking table, which would be needed for such tests, the facility employs a massive reinforced concrete reaction wall about six stories high, where hydraulic actuators are attached to push against test specimens. During the 1980s, successful experiments were conducted on the full-scale testing of a six-story reinforced concrete-frame building and later a matching six-story steel-frame building under a joint U.S./Japan cooperative research program sponsored by the NSF and Japan's counterpart scientific unit. It was also this laboratory that was used in the testing of exterior cladding materials on the six-story steel-frame building, as discussed in Chapter 6.

Figure 10-1 Testing scaled model of an eight-story reinforced concrete-frame structure on 20 × 20 ft earthquake simulation shaking table. *Source:* Earthquake Engineering Research Center (EERC), University of California at Berkeley.

After experimental research in testing laboratories, the next step is implementation. This is where research results are put into practice to reduce earthquake hazards through respective mitigation programs. Many research results, after successfully passing rigorous testing, find themselves incorporated into the seismic provisions of the building code. Others may become part of the fundamental basis developed for the seismic strengthening and rehabilitation of older, existing hazardous buildings that are not retroactively cov-

Figure 10-2 Experimental testing of a base isolation foundation pad. *Source:* Earthquake Engineering Research Center (EERC), University of California at Berkeley.

ered by the building code. In any event, after implementation, many experimental research results may be clearly identified as having made a direct contribution to earthquake hazards mitigation programs. One of the most recent research efforts in 1990 involves the development of "intelligent" buildings or "controlled" structures that contain active mass dampers and electronic sensors that, when activated by excessive motion, trigger internal counter forces against earthquake-induced loads.

ROLE OF BUILDING CODES IN EARTHQUAKE HAZARDS MITIGATION

Historically, building codes have been one of the principal mechanisms used to reduce the effects of earthquakes. The philosophy is that as building code provisions improve, so will the seismic performance of buildings. In general, this has proven to be true. Over the years the records indicate that modern buildings designed to be earthquake-resistant have performed rather well. There are exceptions to this of course, mainly due to cases where the building was not built according to the architect's construction drawings and/or when poor construction methods or materials were used on the site. But in the meantime, it is clear that consistently less and less damage occurs to well-designed structures in areas of high seismic risk.

In his comparison of earthquake codes, Rene Luft (1989) indicates that earthquake design provisions on a national basis are contained in documents published by four organizations:

1. "Minimum Design Loads for Buildings and Other Structures," ANSI Standard A58.1-1982, American National Standards Institute.
2. *NEHRP Recommended Provisions for the Development of Seismic Regulations for New Buildings,* Federal Emergency Management Agency, 1985 and 1988 editions (NEHRP: National Earthquake Hazards Reduction Program);
3. "Recommended Lateral Force Requirements and Tentative Commentary," *Blue Book,* Seismology Committee, Structural Engineers Association of California (SEAOC), June 1988; and
4. Uniform Building Code (UBC), International Committee Conference of Building Officials, 1985 and 1988 editions.

His review of the respective provisions indicates that some of the essential differences between these requirements are:

(a) The NEHRP document gives force levels corresponding to a strength-based or limit states design, while the other three documents give force levels that correspond to working or service stress designs;
(b) the importance factor is used as a multiplier of the base shear level in all documents except NEHRP, which treats building importance by a hazard exposure group;
(c) NEHRP and UBC-1988 contain detailing requirements for all common construction materials and all seismic zones, while the UBC-1985 contains detailing requirements for zones of high seismicity but only limited requirements for zones of moderate seismicity;
(d) P-delta analysis is specified by NEHRP for all buildings that must be analyzed, by SEAOC for buildings that exceed drift limits, by UBC-1988 for all buildings except those in Zones 3 and 4 meeting drift limits, and is not specified by ANSI.

In reviewing all four documents, it is apparent that the UBC is a building code, and that ANSI, NEHRP, and SEAOC are reference documents. The SEAOC provisions are mainly for application in California. NEHRP is the specific document "developed as a source document for use in all earthquake-prone areas in the United States."

California Field Act of 1933

In California, as discussed in Chapter 8, in order to mitigate earthquake hazards, the first set of seismic provisions to be developed were placed in a special section of the building code after the 1933 earthquake in Long Beach. The goal of the building code was to reduce the effects of earthquake hazards by improving the seismic performance standards of buildings. But perhaps even more important was the fact that at the same time, the California Field Act was enacted into law by the state legislature.

California Assemblyman C. Don Field witnessed the collapse of many structures in the 1933 Long Beach earthquake. Many were public school facilities, and because seismic performance standards for any building type did not exist at the time, they were not designed to withstand earthquakes. Within a month after the earthquake, a bill was sponsored by Field, and enacted, that required all new public schools to be designed and constructed to resist earthquake-generated forces. The basic philosophy was that since California required attendance in a specific school structure until the age of 16, and in so doing forced students to accept an "involuntary risk," the state was responsible for their safety. It was agreed that the least that could be done was to place students into a safe setting.

Furthermore, the new law mandated that the construction drawings for any new public school had to be plan-checked and approved by the Office of the State Architect rather than by a building department at the city or county level. This provision was deemed necessary at the time as a quality-control program to guarantee that the same design and construction performance standards would be uniformly adhered to throughout the state. What the Field Act did at the time was to set a precedent that singled out public schools and required that they be planned, designed, and constructed under a quality-control process, governed by the Office of the State Architect, that ensured higher performance standards than those for other building types in the state.

The Field Act is still recognized as having contributed substantially to hazards mitigation in the annals of U.S. earthquake preparedness. California now has a 56-year history of earthquake-resistant public schools. Again, since the Field Act was not retroactive, public school buildings existing prior to 1933 were not covered. Consequently, a follow-up act was passed in the early 1960s that required that all existing, hazardous pre-1933 public school buildings be abandoned or undergo a comprehensive seismic retrofit by the year 1977. Although the deadline was extended to 1978 to accommodate a few local public school districts placed in distress by the law, California can now boast that not even one earthquake-hazardous pre-1933 public school building is in use. In conclusion, it must be indicated that the provisions of the Field Act have been justified over the years, as public schools in California designed and built since 1933 have performed well during earthquakes, of which there have been many. Even in the October 1989 Loma Prieta earthquake, damage to public schools was minimal.

The California Hospital Act of 1972

Much later in California, using the precedent established by the Field Act, another earthquake hazards mitigation effort was enacted after the 1971 San Fernando earthquake, in which the acute vulnerability of medical facilities to earthquakes was revealed. The Hospital Act of 1972 was passed for the public safety of hospitals. As explained earlier in Chapter 8, whereas the Field Act requires that school buildings withstand an earthquake without collapse or injury to their occupants, the Hospital Act requires that emergency or critical facility buildings must not only remain standing but also

Figure 10-3 New hospital building in California designed and built under 1972 Hospital Act.

"functional" or "operational." While the Field Act did not address damage to nonstructural components and/or contents, the Hospital Act plainly does. By doing so, it unquestionably raised the reduction of earthquake hazards to an additional, higher level. (See Figure 10-3.) No critical damage to hospitals designed under the new performance standards legislated by the California Hospital Act was reported after the October 1989 Loma Prieta earthquake.

At the international level, for obvious reasons, the highest hazards mitigation standards enacted for any potential disaster, including earthquakes, are required in the planning and design of nuclear power facilities. Accordingly and rightfully, if nuclear facilities continue to be considered an alternative energy source around the world, the highest priority must be given to establishing performance standards for their planning, design, and construction.

PREPAREDNESS PLANNING

Earlier it was mentioned that the principal purpose of earthquake hazards mitigation programs was to undertake efforts to reduce the impact of a potential disaster before the event rather than after. This is the exact role that preparedness planning plays: to identify, develop, and implement programs that will alleviate the impact of a future earthquake before it occurs. One might even define it as being "anticipatory" planning rather than postevent "recovery" planning. Much grief and great losses can be avoided if some aspects of hazards mitigation planning can be instituted in anticipation of the event by recalling that it is not a question of if the earthquake will occur, but when. From the architectural planning point of view, we are clearly talking about preparedness efforts. And, within this area many steps can be taken.

Barriers to Mitigation and Preparedness

Lack of earthquake hazards mitigation programs and preparedness plans are commonly due to a community's failure to grasp the potential impacts of seismic activity in the area. By misunderstanding the probable recurrence intervals of earthquakes, say one that is said to have a 100-year return period or that is beyond the life expectancy of the average person, there is a tendency to rationalize that "one will not occur during my lifetime." Such rationalization ignores the possibility that two 100-year-recurrence earthquakes may occur within the same year or two, as so frequently happens in the case of the so-called 100-year flood or another type of natural disaster. There is a failure on the part of the public not to personalize the consequences of an actual event when the probability of its occurrence is not an immediate threat.

In addition there is the "Act of God" syndrome, which includes the suggestion that nothing can be done to stop it, so why bother? This echoes the feeling that a disastrous event that is about to happen is so big that nothing can be done to help—we are all doomed and in the hands of God anyway, so let nature take its course. On local government's part, there is always the excuse that there aren't enough resources or funds available for earthquake hazards mitigation. Consequently, there are always more important things to worry about. And finally, when there is a lack of data or inadequate information on the subject, confusion results and preparedness efforts lag. Of course, after the damaging earthquake, all these excuses are forgotten, and at that point the community has a tendency to ask, "Why wasn't something done 10 years ago to prevent this?"

As can be seen by the above reactions, one of the main keys to preparedness is found in education. In many communities that have successful preparedness programs, television has been used effectively as a mass communications device to present educational programs geared to various age groups. Additionally, information is disseminated starting with educative classes and drills in public schools, and there are also adult education programs available. The goals are to structure attitudinal changes.

One of the main purposes of the three regional earthquake preparedness projects in the United States, BAREPP, SCEPP, and CUSEPP, is to supply educational materials openly and freely to the public and private sectors of a community. The need to raise public awareness of the earthquake problem has been identified as one of the first targets. Once this attitudinal change has occurred, it is believed that "half of the battle will have been won."

HAZARDOUS BUILDINGS ABATEMENT

As indicated in Chapter 7 and other sections of the text, unreinforced masonry buildings are ranked high on the list of potentially hazardous structures. Their poor performance in the 1933 Long Beach earthquake was responsible for the Field Act, as indicated above, that only targeted public school buildings. This raises the following question: What about all of the other types of old unreinforced masonry buildings that are still in use all over the United States, especially in the Midwest and on the East Coast, and around the world?

Even in California, the home of the Field Act, the greater Los Angeles basin region alone has approximately 20,000 pre-1934 unreinforced masonry buildings, while about another 8000 exist in San Francisco. As the chief building official of the city of Palo Alto indicated, "If unreinforced masonry buildings were on wheels, they'd be recalled." One potential earthquake hazards mitigation project, where success in partially reducing the damaging effects of a seismic event is assured, is to develop a hazardous building abatement ordinance.

Many sectors of the United States are less receptive than California to the retrofit and rehabilitation of existing hazardous structures, so the problem of hazardous building abatement must be addressed in digestible stages. This even applies to California, where an immediate wholesale solution to the problem, if even feasible, would certainly have disruptive economic consequences, as more than 50,000 pre-1934 buildings are estimated to exist in the state.

Consequently, a new law was passed in 1986, Senate bill 547, that introduced an initial step in redressing the problem. The preparatory program required only those cities and counties in Seismic Zone 4 to identify and quantify the numbers of unreinforced masonry buildings in their jurisdictions. Upon completion of the survey, each jurisdiction had to develop a program to define the magnitude of the problem and how to deal with it. No further action was required until the program objectives and methods in dealing with the problem were approved. To guarantee rapid compliance with the new state-mandated law, all evaluation studies and the final report of recommended building abatement requirements had to be completed within four years, by January 1, 1990. In the final report, each community had to establish a mitigation plan and be prepared to implement it immediately. Despite the strict deadlines given, the program is reportedly behind schedule in several small local jurisdictions.

EARTHQUAKE PREDICTION

A growing constituency sees the ultimate approach to earthquake hazards mitigation resting on the success of predicting earthquakes. This group believes that an appropriate warning process based on accurate forecasting is the only solution to the earthquake challenge. Clearly, the possibility of predicting seismic events adds a new dimension to earthquake-preparedness and disaster-response planning. However, it must be understood that even when earthquake prediction becomes flawless, buildings still must be designed to resist earthquake loads. Predicted or not, earthquakes will continue to damage any structure not properly designed. An effective method developed to predict earthquakes will certainly save lives, but not necessarily all buildings.

Funds supporting earthquake prediction research have increased substantially on a worldwide basis during the last 25 years. In the United States over $17 million per year is spent on earthquake prediction research, while other substantial programs are also in progress in Japan, the USSR, and China. However, as yet, methods thought to be effective at the time have not been replicated successfully—the successful prediction of the Haicheng, China, earthquake in 1975 was followed in 1976 by the totally unexpected, unpredicted, earthquake in Tangshan, China, which killed 250,000 people.

The ultimate goal of prediction is the ability to forecast earthquakes scientifically and accurately, which means knowing their exact (1) time and date of occurrence, (2) location, and (3) magnitude, well before they occur. Why is this so important? In terms of public health and safety, the principal objectives of a credible earthquake prediction technique are to (1) save lives and preclude injuries, (2) curtail property losses, (3) avoid community disruption, and (4) alleviate ancillary economic and social effects. All the interest in earthquake prediction centers on the fact that a successful prediction would offer an unprecedented opportunity to save lives and property from destruction. It is often said in research circles that the scientist who develops an accurate method to predict earthquakes, one that can be replicated, will be awarded a Noble Prize hands down, without question.

In attempting to develop a scientific model for predicting earthquakes, researchers are striving to identify and record specific precursors that may signal an imminent earthquake on either a short-term or long-term basis. Precursors being examined by these scientists include the following, among others:

1. Anomalies in the rate of strain accumulation along the fault,
2. Unusual regional subsidence or uplift mechanisms,
3. Sudden changes in groundwater levels,
4. Variations in local magnetic fields (geomagnetism),
5. Increased release of underground gases, such as radon,
6. Changes in heat flow from the center of the earth,
7. Measurable shifts in ground tilting,
8. Appearances of seismic gaps between sections of a fault where previous earthquakes occurred, and
9. Uncustomary animal behavior.

The seismic gap theory is based on the possibility and perception that an earthquake is more apt to occur along segments of active faults where a long period of time has elapsed since the last seismic event. Although the plotting of all historic earthquakes along a specific active fault is relatively easy, pinpointing the precise spot, time, and size of the event is not. The 1985 Mexico earthquake filled a recognized seismic gap, and is said to have been the seventeenth major earthquake to be identified as fitting into the seismic gap theory. It has also been said that the October 1989 Loma Prieta earthquake filled in a seismic gap in northern California which existed along the San Andreas Fault in the Santa Cruz Mountains.

Although several U.S. scientists are intrigued but somewhat doubtful of the conclusiveness of studies on animal behavior before an earthquake, some researchers in China and Japan are still seriously conducting such studies. Attention to animal behavior before an earthquake is nothing new in Japan, as can be witnessed by some of its popular mythology; there was even belief that a legendary animal precipitated the event. Several historic reports give detailed accounts of how the Great Catfish caused earthquakes by surfacing to the top of the sea instead of resting casually at its bottom (refer to Chapter 1 and Figure 10-4). Yet, while in China on a recent trip, when the author, in view of the successful prediction of the 1975 Haicheng earthquake followed by the unpredicted Tangshan earthquake disaster in 1976, asked an official in the Office of Earthquake Resistance in Beijing what he personally thought of monitoring animal behavior as a precursor of earthquakes, the following curt response was received: "It's for the birds!"

The subject of earthquake predictions often conjures up images related to psychics, seers, astrologers, animal behavior, religious prophets, and wild scientists. Yet, many reliable professionals have

Figure 10-4 Japanese print showing Giant Catfish (Namazu) pinned down by Kashima deity. *Source:* Reprinted with permission of Teledyne Geotech; Garland, Texas; 1985.

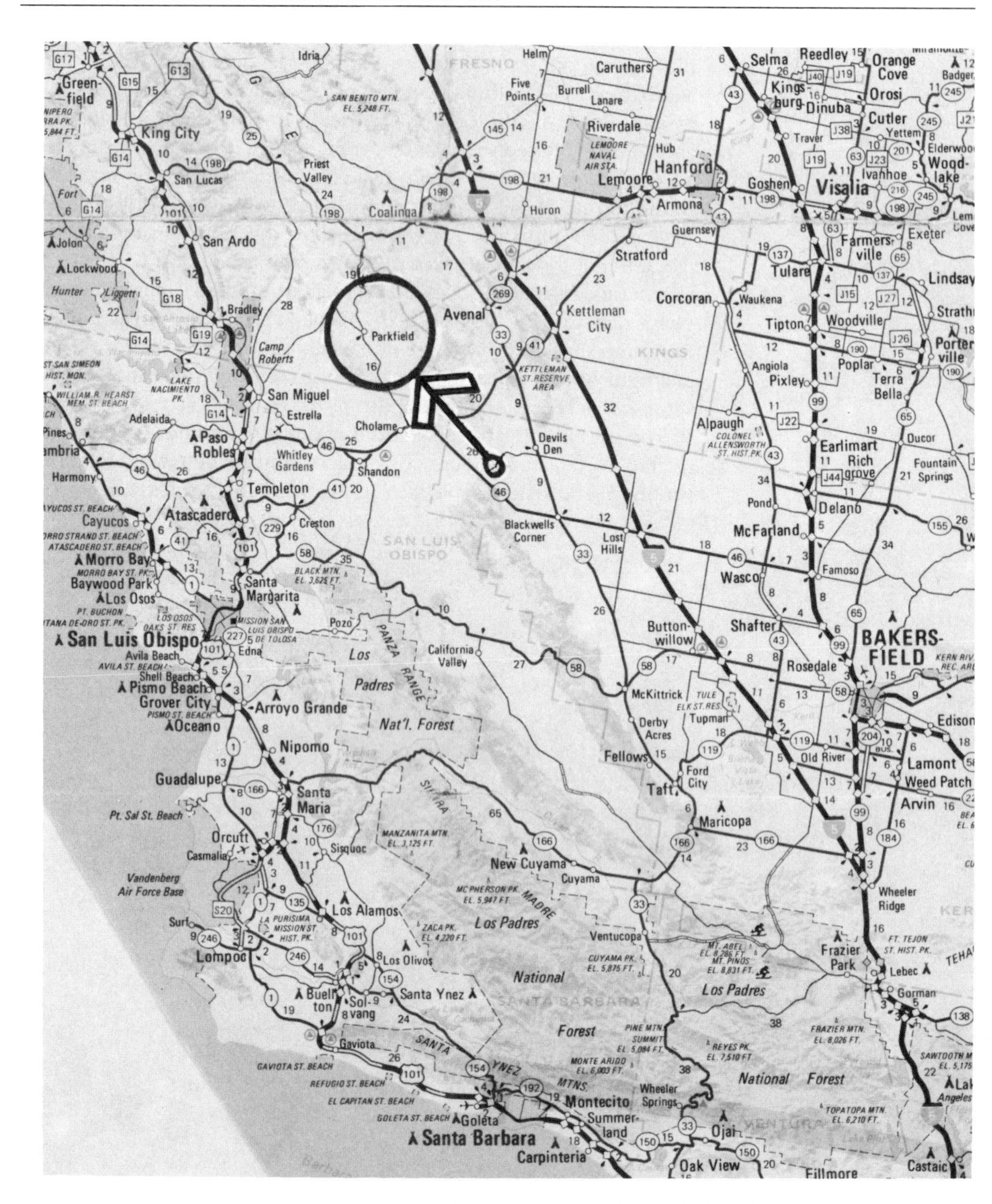

Figure 10-5 Map showing location of Parkfield in relationship to Pacific Coast of California, Santa Barbara, San Luis Obispo, Fresno, and Bakersfield. (Note location of Coalinga, site of the 1983 earthquake, immediately above Parkfield.)

been and still are involved in earthquake prediction at one time or another: geologists, seismologists, geotechnicians, mineralogists, paleontologists, zoologists, and botanists. In the United States, a major effort to develop a scientific formula for successful prediction is under way right now: the Parkfield Earthquake Prediction Experiment. It will attempt to make a short-term prediction of the next earthquake on the San Andreas Fault in central California, where historic records indicate that a moderate earthquake occurs regularly on an approximate 20-year cycle. As the last earthquake in the Parkfield zone occurred about 24 years ago, in 1966, instrumentation has been increased in the immediate area in an effort to identify and record precursors to a moderate earthquake. (See Figure 10-5.) U.S. Geological Survey (USGS) scientists have assigned a 95 percent probability to the likelihood that a moderate 6.0 magnitude earthquake will occur along the fault in the Parkfield zone. If successful, for the first time specific precursors may be accurately recorded for further analysis and development.

Relationship to Preparedness Planning

If the Parkfield Experiment is successful, it will also allow response planners to test their emergency plans on a realistic basis. However, this raises the following generic questions: How will the public be informed and how do we know that it is a valid prediction? If the prediction is validated scientifically, what will preclude wholesale panic and mass exodus from the area targeted by the prediction? All freeways and airports would be bedlam. If, for public safety reasons evacuation from hazardous central business areas is recommended and commercial activities and banking transactions are suspended, how will the ensuing economic losses be recovered? What special preparations should be taken for schools, medical facilities, prisons, and airports? How will state and local government be compensated for extra costs, and how will preparedness costs be financed?

This is where the timing of a prediction before an earthquake is important, as it is hoped that enough lead time will given for appropriate preparations. Hopefully, it will not be a matter of seconds, since for adequate preparation, the longer the lead time the better. An official plan already should be in place to (1) receive the prediction, (2) evaluate it, (3) announce it to the public, and (4) prepare for the earthquake.

In the United States, receipt and validation of a scientific prediction would take place at two government levels, federal and state. At the federal level, the director of USGS, Department of the Interior, has been given the responsibility of acknowledging a bona fide earthquake prediction and conducting an evaluation of its scientific merit on a national level. However, his office is not required or expected to take any response actions other than indicating the validity of the prediction and issuing an earthquake "alert." Response and preparedness plans are the responsibility of state government.

It is clear that all states, especially the more earthquake-prone states on the West Coast, in the Midwest, and in the Northeast, must develop detailed plans on the manner in which predictions are officially received, given consideration, evaluated, and announced. In view of the Parkfield Experiment, California's state government has set up an elaborate procedure to manage the situation. The need for this kind of management plan was illustrated by prior experiences subsequent to the release of three separate prediction "advisories" following what were judged to be possible precursory events in San Diego (1985), Chalfant Valley (1986), and Lake Elsman (1988).

Using California as a model, the following structure has been established: the governor's Office of Emergency Preparedness (OES) has the primary role in receiving earthquake predictions, organizing official administrative actions, and directing subsequent operations at the state level. In 1985 the OES was given the responsibility to "develop a comprehensive emergency response plan for short-term earthquake predictions." This office is supported by the geophysical monitoring activities of the California Division of Mines and Geology (CDMG) and technical advising by the California Earthquake Prediction Evaluation Council (CEPEC). The council is composed of leading earth scientists throughout the state who have the authority to consult with USGS and review, assess, and validate the data on which the prediction has been made. CEPEC assists the Office of Emergency Services and the governor to determine the scientific and technical merit of the prediction.

The Role of the Architect and Definition of Terms

Even though scientists and researchers indicate that the development of a truly reliable earthquake prediction process may not be realized for many years, architects still need to be familiar with the technical terms used in earthquake prediction preparedness plans. It is quite apparent at this point that increased apprehensions and anxieties by clients and building owners who hear about a prediction will probably prompt calls to their architects for recommendations on mitigation actions, if any, to be taken. Some of the more commonly used terms are from Wallace, R. E., et al (1984)

Long-term Prediction: A prediction of an earthquake that is expected to occur within a few years up to a few decades.

Intermediate-term Prediction: A prediction of an earthquake that is expected to occur within a period of a few weeks to a few years.

Short-term Prediction: A prediction of an earthquake that is expected to occur within a few hours to a few weeks.

Alert: A forecasting term used as a subdivision of a short-term prediction applied to the period of three days to a few weeks.

Imminent Alert: A forecasting term used as a subdivision of a short-term prediction applied to a period up to three days.

Advisory: A formal forecasting message giving earthquake information or advice to take action.

To refer back to the Parkfield Experiment, a number of sensitive measuring devices were installed in the area to monitor any signals that would indicate that changes were taking place as precursors to a seismic event. This instrumentation includes seismographs, creepmeters, strainmeters, extensometers, well-water-level measuring devices, laser arrays, and magnetometers. When sufficient changes in the data collected by the instrumentation are interpreted as indicating an increased likelihood for an imminent earthquake, the USGS will issue an alert. When "alert level A," which indicates a 37 percent or greater probability for a magnitude 6.0 magnitude earthquake occurring in 72 hours is reached, the OES will implement its Parkfield prediction response plan in which OES offices in the seven counties will be informed of the level A alert. At this juncture the counties are responsible for making certain that all emergency responders within their jurisdictions are ready for action. All components of mass media outlets, radio, TV stations, and newspapers, will be notified to provide the general public with details of the prediction and advice on what to do.

The bottom line of all this for architects is that they must be fully aware of the ramifications of an earthquake prediction in order to be prepared to offer advice to their clients or building owners, if required and/or asked, on which appropriate mitigation activities could be taken after a technical forecast or earthquake advisory has been issued. Unless architects are familiar with the social and economic impacts of earthquake prediction, they will be unprepared to suggest appropriate actions when an advisory is received. At this point, architects must understand that, given the state of earthquake prediction efforts, short-term advisories will be issued in the future—of this, there is no doubt. As professionals, they must be in a position to treat it seriously, advise their clients when necessary, and be ready to take appropriate actions in preparing for an earthquake.

CHAPTER

11 RECOMMENDATIONS AND SUMMARY

Codes and ordinances concerning earthquake-resistant measures have limitations in their contents, the most obvious one being that the written word can never take the place of the design process and the actual physical act of construction. That part rests with the architect, who is trained to translate ideas, words, design methods, and graphics into an actual physical expression embodied by coordinated building systems in harmony with the natural environment. As earthquakes are a manifestation of that natural environment, it is advantageous for architects as responsible design professionals to understand the implementation process in earthquake hazards reduction procedures just as thoroughly as they understand those for design programming, energy conservation, handicapped access barriers, environmental pollution, building performance standards, and economic constraints.

THE CHANGING SCENE

By now it must also be clear to architects that earthquake-resistant design is still considered a state-of-the-art process, one that is still striving to develop a finite scientific format that is absolute and free of uncertainties. Building code standards change repeatedly and revisions to the Uniform Building Code take place every year, with a new edition published every two to three years. Seismic provisions in building codes are not exceptions to this norm; they also change quite frequently, particularly after a major damaging earthquake that takes everyone by surprise (recently in Mexico in 1985, in Armenia in 1988, and in the Santa Cruz Mountains of California in October 1989).

As public awareness of seismic safety concerns increases each year, architects must keep themselves fully informed of new seismic developments at all times if they are to compete successfully in the marketplace, nationally and internationally. Many universities around the country are constantly engaged in research activity through organized research units to unravel the mysteries of earthquakes and their causes and effects. These research units, as public institutions of learning, will gladly furnish information on their latest research findings to anyone interested enough to give them a call, whether it be on base isolation, internal energy absorption components, the nature of the forces driving the motion of plates around the earth's surface, advances in earthquake prediction, or whatever. It behooves the architect to know where these institutions are, and how to obtain the latest information from them. The constant reading of new technical books, study of the latest research papers, attendance of professional continuing education seminars, participation in professional conferences and symposia, and a basic understanding of an emerging body of knowledge are all necessary to keep ahead of changing design methods, construction practices, performance standards, and earthquake hazards mitigation procedures. The dimensions of seismic design must be extended beyond the realm of structural considerations alone.

ARCHITECTURAL ROLES IN SEISMIC SAFETY

The message of this book, in case it isn't yet clear, is that the design architect must become thoroughly immersed in an earthquake hazards mitigation role from the very beginning of the site development and design process or face a rude awakening when the next major earthquake hits. At that time, the architect will be put on the spot if building damage patterns are severe and collapses have taken place. (See Figure 11-1.) The structural engineering profession has a well-established history and an impressive record in seismic research, design, and implementation of earthquake hazards reduction measures, all of which have been openly publicized. The profession has obviously been very active in seeking advancements in earthquake engineering theory and practice. The architectural profession as a whole has not as yet taken up the challenge.

In fact, many design architects continue to have dismaying attitudes concerning earthquakes: "It won't happen here" (which it will); "It's strictly an engineering problem, so my structural consultant will take care of it" (which it is not); "It will cost too much" (which it will not); or "The earthquake threat has been overblown and exaggerated by a bunch of doomsdayers" (which it has not). (See Figure 11-2.) In this day and age, any architect who continues to harbor such naive thinking had better be very careful in public, particularly after an earthquake such as the 7.1 October 1989 earthquake that struck the residents of the Marina District in San Francisco Bay or the drivers of automobiles on the I-880 Cypress Street Overpass in Oakland (see Chapter 12). The architectural profession as a whole can no longer afford to hold such attitudes. Too many advances have taken place during the last few years for the architect to be lulled into believing that it is still just an engineering problem. The design architect must not be complacent and continue to take refuge by hiding behind the consulting engineer. After the next major earthquake in a U.S. metropolitan center, the public and government officials, as we already know from recent experiences in Mexico City, Armenia, and northern California, will certainly attempt to seek out a scapegoat. The architectural profession at this stage needs to make sure that it is not so identified. (See Figure 11-3.)

Table 11-1 indicates areas of potential responses to be assumed by someone during the specific time periods associated with earthquakes. To avoid being fingered as the scapegoat after the next major damaging earthquake, the architectural profession must take an appropriate professional involvement at each level shown. There are many opportunities in these time periods for participation by architects.

In the final analysis, the individual architect who will win out is the one with the capacity to interpret emerging bodies of knowl-

Figure 11-1 Collapsed five-story reinforced concrete-frame building, 1972 Managua, Nicaragua, earthquake.

Figure 11-2 Severely damaged building designed with many offsets in configuration, 1972 Managua, Nicaragua, earthquake.

edge and put them to work. It is strongly recommended that the architect, as a design professional working in coordination with the other planning and design professions, identify fundamental goals and establish specific seismic safety priorities by focusing on the points discussed below.

1. *Implementation of Basic Technical Knowledge and Earthquake Hazards Mitigation Procedures* Architects must know the latest technical developments and be ready to use them when appropriate. It would be unquestionably poor practice not to use the numerous advances being made in earthquake-resistant design methods advantageously. In the same way, the architect must be thoroughly knowledgeable in the latest recommendations on earthquake hazards mitigation techniques and procedures in providing

Figure 11-3 Earthquake-ravaged Italian hill town, Conza della Campania, 1980 Campania-Basilicata, Italy, earthquake. *Source:* Mader and Lagorio (1987).

for life safety in buildings. Concurrently, the client must be made well aware of the cost benefits in doing so. This is especially true during the preevent phase when there is a tendency to forget about and dismiss the need for mitigation measures and preparedness plans.

2. *Participation in Immediate Postearthquake Emergency Phase* An architect must be prepared to participate in actions required during the emergency period immediately following an earthquake. Not all activities call for the involvement of architects, but there are several that clearly need their participation. Two obvious ones that naturally fit in with the services provided by architects are (1) assistance in preliminary building damage assessments and (2) provision of emergency housing needs. (See Chapter 9.) Local government and building owners need to know as soon as possible which buildings are safe to enter and reoccupy, and immediately following a severe earthquake, there are never enough design professionals in the local area to do the job expeditiously.

In the provision of emergency housing, again, the natural creative talents of architects should be tapped. They should volunteer their services and be ready to contribute their expertise to solving the immediate shortage of housing units. Architects should also be involved in postearthquake reconnaissance field studies, as this is where new lessons are learned.

3. *Participation in Intermediate Postearthquake Recovery Phase* Similarly, architects must be prepared to participate in the

TABLE 11-1 Time Periods of Responses to Earthquake Impacts

Phase	Duration of Time Period	Appropriate Responses
Preevent	Hours to decades	Mitigation programs Preparedness plans Research
Earthquake	4 to 60 sec	Seek safety Help others
Immediate postearthquake emergency	1 to 7 days	Search and rescue Medical relief Food and subsistence Assess damage Emergency housing Emergency planning Temporary lifelines Clear ruble & debris
Recovery	Days to Months	Temporary housing Repair lifelines Evacuation Clear damaged buildings. Emergency Codes
Reconstruction and continued recovery	Months to decades	Restore lifelines Existing building rehabilitation Permanent new housing Open space New construction New codes New planing ordinances Earthquake memorials Mitigation programs[a] Preparedness plans[a] Research[a]

Source: Bay Area Regional Earthquake Preparedness Project, California State Office of Emergency Services.
[a]For next earthquake.

planning and design of temporary housing needs. Their assistance could also be used in the preparation of emergency codes.

4. *Participation in Long-term Postearthquake Reconstruction Phase* New permanent housing, rehabilitation of damaged buildings, new major construction programs, development of new building codes and planning ordinances, provision of well-designed open spaces, urban planning and design strategies, and the design of earthquake memorials, if wanted—all represent opportunities for the involvement of the architectural profession. Contributions by the architect are desperately needed during this long-range period.

5. *Achievement of Public Health and Safety Targets* There is no question that the architect has a responsibility to create more seismically safe environments and avoid the creation of unsafe ones. In conjunction with this, architects must constantly remember that, as stated in the introduction to this publication, "earthquakes don't kill people, buildings do." Clients must understand this also, as the benefits of reduced casualties in buildings and outdoor areas far outweigh most other considerations. In earthquake hazards mitigation efforts, clear priorities must be established, and life safety, which has always been at the top of the list, must continue to remain there.

6. *Reduction of Building and Property Damage Levels and Urban Economic Impacts* The negative impact of earthquakes on urban economics has always diminished available resources and redirected their use from other desirable services beneficial to the community. Earthquake-generated damages, and the resultant massive property losses, cause an absolute waste of funds because a small fraction of those monies properly spent on earthquake hazards reduction measures before the event would have ameliorated the problem. Yes, federal low-cost disaster loans are available, but that is only a result of crisis planning, not an attempt to find a solution to the problem. Design professionals deal with detailed budgets and property values all the time, so the architect should lobby to make this waste of resources understood by the public and by government administrations. By understanding cost-benefit analysis, architects need to make their clients re-aware of that old adage, "Sometimes a penny well spent is better than a penny ill spared."

In several cases, some of our metropolitan centers located in areas of high seismic risk are "disasters waiting to happen" because the probability of an earthquake recurrence is ignored. An excellent example of this is given by San Francisco, where much life loss, damage, and suffering took place in the Marina District during the October 1989 Loma Prieta earthquake. Seismicity maps of the United States clearly pinpoint questionable locations. In addition, many maps carefully delineate an area of poor soils subject to liquefaction or subsidence during severe ground shaking. Architects know this, and are in an enviable position to do something about it if they wish.

7. *Selection of the Best Design Team Possible* Repeatedly, the architect is the one professional who has an overall perspective of the entire building program, design, and construction process. Architects are trained to synthesize the many variables found in the planning and design process and model them into an appropriate solution that meets the client's needs. The architect represents the client, prepares the contract documents, appoints the structural engineer and other professional consultants, controls the quality of the job, supervises the general contractor, and programs the sequences of the project to facilitate a quality performance and obtain a superior product. Architects, therefore, are in a central position to influence the planning, design, and construction of factors conducive to attainment of seismic safety goals. However, because of the complexities involved in earthquake-resistant design, the architect cannot "go it alone." Earthquake-resistant design demands expertise in various technical fields, as reported earlier, and requires a comprehensive, balanced team approach. Therefore, when it comes to the design of a major project, the architect must select the best team possible to do the job and advise the client that this is the way it must be done.

8. *Assessment of Site Conditions and Secondary Collateral Effects* Appropriate approaches to the earthquake-resistant design of buildings start with the site. Again, here there is enough information available for the architect to get a pretty good fix on the attributes and seismic deficiencies of a given site. In doing so, architects must develop their instincts to look not only at the site under assessment but also at adjacent sites and surrounding areas for potential collateral effects, such as dam failures upstream, massive sea wave run-up from another direction, or a substantial earthquake-induced landslide coming from above. (See Figure 11-4.) In seismic safety site evaluations, it is necessary to take a comprehensive and global view of the situation rather than one only limited to a site-specific examination. Historical air photos of the region in which the site is located will reveal valuable soil failure information helpful in undertaking a broad assessment.

Geotechnical studies are also of help to the architect. In other words, it is just as important to look under the surface of the site as it is to look above the surface. Microzonation studies of the region should also be consulted by the architect for indication of damage patterns produced by past earthquakes in the area. There are many existing maps, earthquake reports, and research studies available on the subject, and architects should use them to their advantage. The architect is in an excellent position to promote land use policies and practices that meet seismic safety standards, and hopefully will have the wisdom to reject sites designated for major developments that do not.

Figure 11-4 Fault-zone surface rupture, 1979, El Centro, California, earthquake. *Source:* U.S. Geological Survey, University of Colorado.

9. *Participation in Advanced Research and Development Initiatives* The engineering, medical, law, and other scientific professions have all excelled in this, so why should the architectural profession lag behind? And yet it is true that architects have not fully involved themselves as a body in experimental research studies and development.

There will always be room for improvement in the earthquake-resistant design of buildings, so why shouldn't architects use their own talents and creative abilities to engage in research study and research applications? The National Science Foundation (NSF) has had funding available in its Earthquake Hazards Mitigation Program for architectural research since 1976, yet very few architects have taken advantage of it. There are about 54,000 architects with membership in the American Institute of Architects (AIA), but it is doubtful that over 200 of them have engaged in formal research activities directed toward the improvement of seismic safety concerns. Architects must not depend on other professions to do earthquake hazards mitigation research for them, but must develop their own research agenda compatible with their own priorities and not those of another profession.

10. *Involvement in Technology Transfer and the Public Arena* One of the critical areas in the advancement of seismic safety relates to how successful the design professional is in the transfer/communication of highly technical data to the layperson and/or public officials in a language that they can relate to. Community acceptance is paramount to final approval of any seismic safety proposals sponsored by the profession.

In order to generate acceptance of its recommendations on the attainment of seismic safety expectations, the profession must devote more time to public policy concerns by meeting with public officials and members of the legislature. It is in the public arena and political spotlight that returns from any expanded effort will be maximized.

11. *Involvement in Public Education Programs* Public education is said to be one of the most important aspects of attaining earthquake hazards mitigation goals throughout the country. Some architects excell in the art of making public presentations of technical material, and these are the ones who should be actively involved in public education at all levels. Progress in this sector of public service will reap great rewards for the profession. Architects must also be prepared to participate in congressional hearings held on seismic safety issues that relate to the profession.

The profession should also develop new programs of education as well as servicing existing programs. New initiatives are needed where traditional approaches have become stagnated. Innovative methods are imperative to public education needs before any dramatic breakthroughs can be achieved by the profession.

CONCLUSION

In order to clarify both the issues and needs of seismic safety goals, architects must assume a leadership role in the development of demonstration programs to provide new directions befitting architectural concerns, whether related to experimental research or implementation measures. It should never be forgotten that for all practical purposes the architectural profession, more than any other, is responsible in all ways for the planning and design of the total built environment and its substantive implications for public health and safety. Too much rides on this responsibility to allow any part of the built environment to be destroyed by the next earthquake, no matter how moderate or severe it may be.

CHAPTER

12

1989 LOMA PRIETA EARTHQUAKE IN THE SANTA CRUZ MOUNTAINS OF THE SAN FRANCISCO BAY AREA REGION

On Tuesday, October 17, 1989, at 5:04 P.M., an earthquake with a surface-wave magnitude M_s of 7.1 and a Richter magnitude of 7.0 occurred on a section of the San Andreas Fault in the Santa Cruz Mountains of northern California immediately south of the San Francisco Bay Area region (see Figure 12-1). According to U.S. Geological Survey (USGS) sources in Menlo Park, this earthquake is said to have reruptured the southernmost 60-km part of the San Andreas Fault break in 1906. It released an amount of energy equal to about 30 million tons of high explosives, nearly 10 times the total of all bombs used in World War II.

Since the epicenter was located 16 km northeast of Santa Cruz, in a mountainous region in which one of the prominent features is the Loma Prieta Peak (elevation 3608 ft), the USGS officially named it the Loma Prieta earthquake. It is the first earthquake in the United States to occur immediately adjacent to a major metropolitan area since the 1971 San Fernando earthquake immediately adjacent to the Los Angeles Basin in southern California.

Although information on this earthquake is still limited to general preliminary data, its dramatic results were felt to warrant discussion in a separate chapter here. Though brief by necessity, the basic intent of this summary is to present the highlights of the earthquake's impact on a major metropolitan area relative to issues of interest to the architectural profession.

Damage patterns associated with this 7.1 magnitude earthquake, as contrasted to the total or near-total destruction experienced in Armenia, Tangshan, Skopje, and Managua, present excellent oppor-

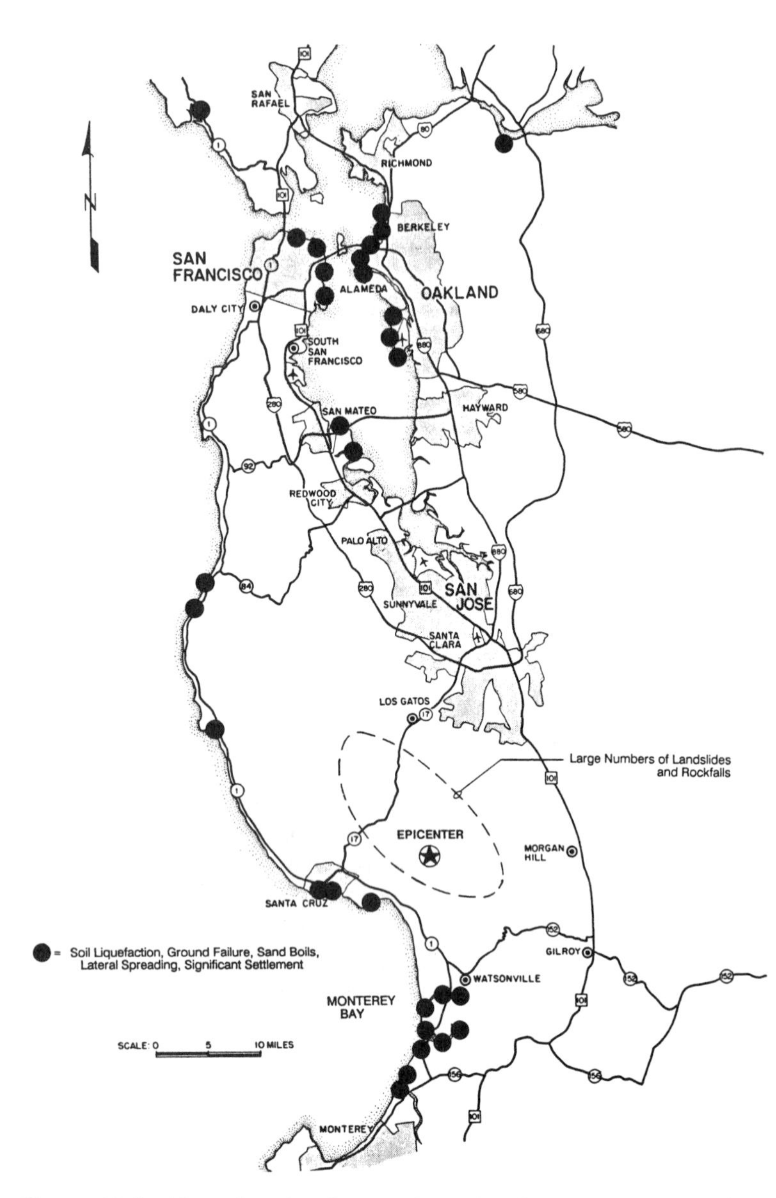

Figure 12-1 Map of region impacted by the October 17 Loma Prieta earthquake. *Source:* University of California at Berkeley. Seed, et al., (1990) Earthquake Engineering Research Center, (EERC).

tunities for the study of architectural and planning seismic issues and concerns.

GENERAL DATA

The earthquake struck on a Tuesday at 5:04 P.M. at the start of the peak evening commute hour (see Figure 12-2). It occurred at the

Figure 12-2 San Francisco Ferry Building Clock Tower stopped at 5:04 P.M. by the October 17 Loma Prieta earthquake.

same time that over 60,000 spectators were sitting in the stands of Candlestick Park waiting for the start of the third game of the World Series baseball game between the Oakland Athletics and the San Francisco Giants. This was a very fortunate situation, since the stadium performed very well during the earthquake, sustaining only minor damage, and consequently there were no casualties or panic among the huge crowd. The implication of this was that 60,000 persons were not in the downtown areas of San Francisco and Oakland at the time of the earthquake, and in particular, not driving on usually congested freeway overpasses and bridges during the rush hour. As will be seen later, this was indeed a fortuitous situation.

The death toll from this earthquake is officially listed as 67 persons, admittedly low for a 7.1 event that occurred at the height of the evening rush hour. As more buildings are carefully inspected and the extent of damage analyzed, property loss estimates could run as high as $10 billion. By October 27, reports indicated that there had been 2435 injuries and about 13,000 rendered homeless (although some later reports indicated a total of approximately 20,000 homeless). As of November 7, there were 13 Red Cross shelters open, with 1202 still listed as homeless.

MAJOR DAMAGE IMPLICATIONS OF LOMA PRIETA EARTHQUAKE

Percentagewise and in general, considering a 7.1 magnitude earthquake, the overall built environment performed quite well in the large nine-county metropolitan area of the San Francisco Bay region in northern California. Preliminary evidence indicates that except for the few pockets of major damage in the older, existing areas of specific cities in the impacted region, there was little to moderate building damage. Many areas reported no accounts of any damage of consequence. The specific pockets of severe damage included the Marina District and the South of Market area (SOMA) in San Francisco, the Cypress Street I-880 two-tier overpass area and downtown area in Oakland, downtown Los Gatos, downtown Santa Cruz, downtown Hollister, and downtown Watsonville. In terms of the greatest overall percentage loss to the existing building stock per city, downtown Watsonville, where 324 buildings were destroyed, suffered the most.

Although relative damage was widespread over a seven-county area from the Monterey and San Benito counties in the south, to the San Francisco and Alameda counties to the north, major damage and property loss varied considerably throughout the region from area to area, and seemed concentrated in those distinct pockets described above. Even though some damage also occurred in the Marin and Contra Costa counties farther north, they were not declared federal disaster areas.

Even though this seismic event was not the expected "**BIG ONE,**" which would be equivalent to a 8.3 magnitude earthquake (as in the 1906 disaster), the relatively good performance of buildings and moderate life loss have resulted in a sense of confidence in construction standards and building code requirements for high-rise buildings. Not one high-rise, of which there were many in Oakland and San Francisco (with some as high as 50 stories), failed catastrophically. However, overall, San Francisco officials still reported that out of the 19 areas of San Francisco City and County, only one (Bernal Heights) had no reports of heavy building damage. A relative realization has set in that the main regional problems experienced in this earthquake were primarily limited to the general overall infrastructure and lifeline systems.

As a result of this earthquake, public attention and mass media interests focused almost exclusively on three dramatic events: (1) severe damage to the structure of the San Francisco–Oakland Bay Bridge, (2) collapse of the Cypress Street double-deck Highway I-880 overpass, and (3) collapse of several three- and four-story condominium structures followed by a serious fire in the San Francisco Marina District. Later, the media focused on the heavily damaged areas in the cities of Santa Cruz, which President Bush visited along with his earthquake reconnaissance trip to the San Francisco Bay region, and Watsonville.

Vulnerability Analysis Studies, Property Loss, and Casualties

Vulnerability analysis studies (see Chapter 8), completed in 1972, 1980, and 1987 for preevent planning purposes, presented anticipated earthquake damage scenarios for the San Francisco Bay Area in the event of a recurrence of a 8.3 magnitude earthquake (as in 1906) on either the San Andreas or Hayward faults located in the region. In these reports, predicted property loss and casualty estimates were much higher. In the Federal Emergency Management Agency (FEMA) report (1980), "An Assessment of the Consequences and Preparations for a Catastrophic California Earthquake," casualty estimates as high as 11,000 deaths and 44,000 injuries were forecast for an 8.3 magnitude earthquake on the northern San Andreas Fault; these statistics were proportionately much higher than the actual loss statistics reported.

Accordingly, the 1989 Loma Prieta earthquake, even though it registered a much lower magnitude, 7.1, provides an excellent opportunity to review, assess, and improve casualty and loss estimation modeling used in these vulnerability studies. In the future, it is crucial to these studies that casualties be carefully correlated with the place of injury, type of injury, building class, and construction type in order to achieve more accurate forecasting for preparedness planning purposes.

The relevance of vulnerability studies to the architectural profession is indicated by the fact that architects were participating members on several such study teams and thus had an important role in the prediction of casualty and damage estimates. It is extremely important that architects continue to take part in the devel-

Figure 12-3 Collapsed portion of the upper deck of the San Francisco–Oakland Bay Bridge. *Source:* Gary Reyes, *Oakland Tribune.*

opment of such earthquake scenarios and assist in preparedness planning activities.

Although by and large many aspects of the damage scenarios turned out to be exactly as predicted in these reports issued by federal and state of California agencies, the Loma Prieta earthquake also produced some surprises. Fortunately, one was that the final casualty figured (67 deaths and 2435 injuries) were not as great as anticipated. Two reasons for this were that the Loma Prieta earthquake had a short relatively duration and a 7.1 magnitude; the scenarios had used a 8.3 magnitude for the postulated earthquake, which was roughly more than 30 times the amount of energy released by the actual October 17 event. Another was that, by coincidence due to the tremendous enthusiasm for this particular World Series, over 60,000 spectators were outdoors at the World Series baseball game at Candlestick Park, or at home watching the game on TV, rather than on the road or in vulnerable parts of urban centers, where much of the damage and most casualties occurred. As it turned out, therefore, it was by sheer luck that the number of people exposed to the earthquake hazards was far less than anticipated in the areas where most of the deaths and injuries occurred.

Another surprise was the collapse of the upper deck of the San Francisco–Oakland Bay Bridge (see Figure 12-3). The pre-earthquake scenarios developed by the vulnerability studies had forecast the closure of the bridge for 72 hours owing to ground failures at the eastern approaches to the bridge where they crossed over man-made artificial fill areas with a high water table underneath, but the bridge structure itself had been expected to perform well, without major damage. It was a complete shock, therefore, when the earthquake-induced forces caused the failure of 2-in. diameter bolts on the vertical supports plus other damage patterns that allowed the upper deck to collapse onto the lower deck. Again, fortunately, only nine deaths were caused by this unexpected collapse, one on the upper deck and eight on the lower deck.

Cypress Street Highway 1-880 Overpass

Of the 67 deaths reported, 41 occurred at the site of the Highway I-880 Cypress Street overpass collapse, which had been constructed in an area characterized by estaurine mud deposits. (See Figure 12-4.) Since the earthquake occurred during the peak evening rush hour, where commuter traffic is normally bumper to bumper on Highway I-880's Cypress Street overpass from 4:30 to 5:30 P.M., death estimates were as high as 200 when the total pancake collapse of the upper deck of the overpass was initially observed by search and rescue squads (see Figure 12-5). Automobiles caught between the upper and lower deck were crushed flat (see Figure 12-6). It was considered a miracle that there were only 41 deaths resulting from the collapse, again revealing that the traffic

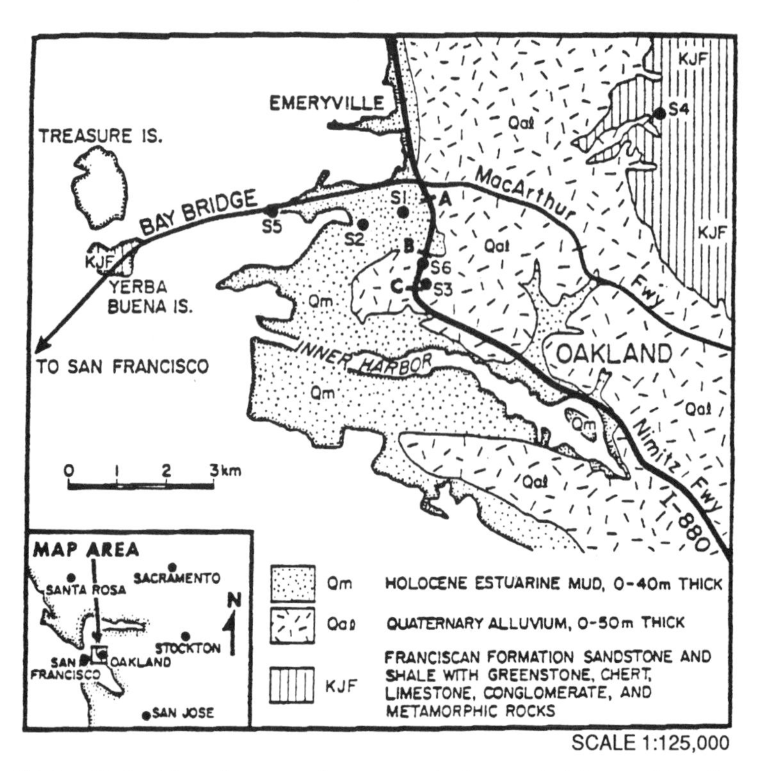

Figure 12-4 Map of near-surface geological characteristics and locations of I-880 Freeway, including collapsed portion (A–B) of the two-tier I-880 Cypress Street overpass. *Source:* Borcherdt et al. (1989), U.S. Geologic Survey.

was very much less than initially thought for 5:04 P.M. during a normal weekday evening rush hour.

The collapse of the two-tier overpass at Cypress Street was attributed to its location in a poor soils area, which was composed of estuarine mud deposits and artificial fills (originally the shoreline of the Bay waters was well beyond the site of the overpass), and obsolete reinforced concrete design standards (the overpass was one of the first two-tier overpasses to be built in the area in 1958). After the 1971 San Fernando earthquake in southern California, the Cypress Street I-880 overpass was scheduled for a complete strengthening effort, as were all highway bridges in the state as part of a statewide highway rehabilitation program, but unfortunately phase two of the Cypress Street overpass program was never completed.

One could well ask the question, What does the architectural profession have to do with this? The answer is direct and simple. A California architect has been appointed by the governor of California as a member of a State Review Board charged to make: (1) an assessment of the seismic performance of the overpass, (2) environmental suggestions regarding its location, and (3) recommendations for its future design and/or replacement. This is a strong indi-

Figure 12-5 Collapsed portion of upper level of the two-tier I-880 Cypress Street overpass. *Source:* Dexter Dong, *Oakland Tribune.*

Figure 12-6 Automobile crushed between upper and lower levels of the two-tier I-880 Cypress Street overpass. *Source:* Paul Miller, *Oakland Tribune.*

cation in itself that architects clearly do have a role in many earthquake engineering considerations.

Regional Transportation Problems

The San Francisco Bay Area is critically dependent on its regional transportation system. This system consists of the Bay Area Rapid Transit (BART) system; four bridges spanning across San Francisco Bay from east to west; the Golden Gate Bridge spanning the entrance to the Bay; ferry boats that travel along several routes but principally from San Francisco to Marin and Solano counties; and several interstate highways that run along both sides of the Bay, primarily in north–south directions.

The four most important links in this regional system were knocked out by the earthquake: (1) the San Francisco–Oakland Bay Bridge, when the upper deck collapsed; (2) Highway I-880, when over a mile-long segment of the top level of the Cypress Street overpass failed (see Figure 12-7); (3) Embarcadero and I-280 freeways in San Francisco, which were heavily damaged and closed to traffic; and (4) Highway 17, which was blocked by massive earthquake-induced landslides in the mountains leading to Santa Cruz. In short, the regional system was heavily crippled immediately after the earthquake. Fortunately, the Golden Gate Bridge, the Richmond–San Rafael Bridge, and the BART system were relatively undamaged and available for use.

For two days normal commute patterns were suspended when workers were asked to stay home to alleviate further disruptions. Essential traffic on both sides of the Bay was required to find alternative routes. During this period, several sections of the heavily traveled highways were practically deserted. Finally, by Friday, October 20, as workers returned to their jobs in San Francisco and elsewhere, traffic started to pick up again. By Monday, October 23, commuters found alternative means of transportation and improvised new routes to reach their destinations.

What solutions were found to this problem? First, again fortunately, the BART system was not significantly damaged. It was available for full capacity by the second and third days following the earthquake after a comprehensive safety inspection of the tube under the Bay, the tunnel into Contra Costa County through the Oakland/Berkeley Hills, and all the stations in the entire system. By Monday, October 23, the BART system added extra cars to handle increased ridership. In the first week of November, BART was averaging 345,000 passengers a day compared to the normal pre-earthquake ridership of about 218,000 daily passengers. A real success story.

Additional transportation was provided by the addition of more ferry boats between the East Bay side and San Francisco. Although initially the ferry boats were not that popular with commuters ow-

Figure 12-7 General view from upper level of the collapsed portion of the two-tier I-880 Cypress Street overpass. *Source:* Craig Riesterer, *Oakland Tribune.*

ing to rainy weather and a 40-minute ride across the Bay (in contrast to BART, which only takes about 8 minutes), by the first week of November ridership increased measureably and averaged about 9300 passengers daily.

Finally, traffic on the remaining bridges increased considerably to make up the difference. From the East Bay side, surface traffic from northern Alameda and Contra Costa counties reached San Francisco by way of the Richmond–San Rafael and Golden Gate bridges. In the South Bay areas, increased traffic flow from southern Alameda and Contra Costa counties to San Mateo and San Francisco was served by the Dumbarton and San Mateo bridges.

In the meantime, repairs to the San Francisco–Oakland Bay Bridge were given top priority status by the state. Repair work proceeded on a 24-hour basis to get this critical traffic link back into operation as soon as possible. Original estimates were that repairs would take a minimum of two months to complete before the bridge could be reopened to traffic. Fortunately, the month of November, normally the start of the rainy season in the Bay Area, turned out to be a dry month, with no notable rain storms, which allowed reconstruction of the upper deck to proceed without serious interruption. By taking advantage of the good weather, repairs went very smoothly and the bridge was reopened to full use on Saturday, November 18, a full month ahead of schedule, which everyone acknowledged was a remarkable feat.

Except for the two-tier I-880 Cypress Street overpass, which is scheduled for complete demolition, the Embarcadero Freeway, and parts of I-280 in San Francisco, which are still closed to traffic, all other routes, including Highway 17 over the Santa Cruz Mountains, were back in operation by mid-November.

Survey, Assessment, and Posting of Damaged Buildings

Chapter 9, "Recovery and Reconstruction," indicated that one activity that must be accomplished immediately after a severe earthquake is "a comprehensive assessment of building damage." Building officials, residents, and owners of buildings in the area must have a technical assessment that identifies which buildings can be safely reoccupied and which buildings cannot. The Loma Prieta earthquake was no exception to this process.

Because of the magnitude of the task, many design professionals were called upon throughout California to join inspection teams and assess the safety of damaged buildings in the six-county area. Literally hundreds of structural and civil engineers, architects (including the author), and building inspectors converged into the earthquake-impacted area to assist in the inspection and posting of buildings. Again note that architects have a role to play in the inspection of certain building classes and construction types (but not necessarily all) during this phase of postearthquake recovery efforts. However, it is very important that architects become members of a team in such recovery activities.

Following a major earthquake, because of life safety reasons and liability exposure, in no way should an architect volunteer to do this mass inspection work alone. In postdisaster work, even in the case of simple one-story structures, it is prudent for architects not to work by themselves. They must: (1) have had considerable previous experience or have completed a postearthquake building inspection training course; (2) work together with a team; (3) be authorized to do building damage surveys by local authorities; and (4) realize that the actual posting of the buildings (i.e., placement of damage assessment placards on structures) is to be done by team members from the local building department and/or department of public works.

It is essential to remember that such volunteers are authorized to conduct these inspections only under the auspices of the local authorities. In most cases after the Loma Prieta earthquake, these assessments of damaged buildings typically took place in three

UNSAFE

DO NOT ENTER OR OCCUPY

Warning:
This structure has been seriously damaged and is unsafe. Do not enter. Entry may result in death or injury.

Comments:

Facility Name and Address:

Date ____________
Time ____________

This facility was inspected under emergency conditions for:

____
(Jurisdiction)

on the date and time noted.

Inspector ID/Agency:

Do Not Remove this Placard until Authorized by Governing Authority.

Figure 12-8 Red placard used in posting of unsafe buildings. *Source:* Christopher Rojahn, Applied Technology Council (ATC).

phases of inspection in which ATC-20 Forms developed in 1989, only three months before the earthquake, were used.

For the first round of inspections, ATC-20 Forms for the *rapid visual inspection* of buildings were used, as well as the red, yellow, and green placards for the posting of damaged buildings (see Figures 9-3 and 12-8). This first rapid visual inspection is intended to take only 5 to 10 minutes per building.

The second phase consisted of a more detailed inspection taking anywhere from 15 to 30 minutes. This was a follow-up to confirm initial impressions of the first inspection or modify them, as the case may be. Finally the third phase, if necessary, consists of a very detailed inspection and analysis of the damaged building to guarantee the building's structural integrity and its ultimate safety. This third phase of inspection could even entail a structural computer analysis and evaluation that could take days to complete. Building owners have the right to appeal the results of any inspection and be able to run an independent review of their own.

Building Location and Poor Soils Areas

The San Francisco Bay Area is well noted for estuarine mud deposits and artificial fill areas that have altered original bay margins, streambeds, and marshlands (see Figure 12-9). As indicated previously, the consistency of soils in these unconsolidated fill areas amplifies seismic ground motions. Such locations are also susceptible to liquefaction in high water-table areas, where the soils are

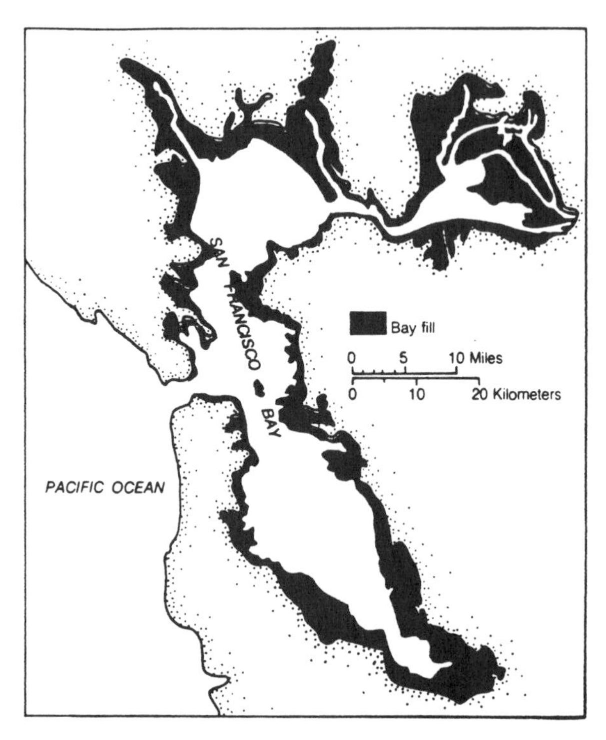

Figure 12-9 Historic Bay margins of the San Francisco Bay Area. *Source:* William E. Gale, Jr.

Figure 12-10 Liquefaction along 7th Street in city of Oakland, 1989 Loma Prieta earthquake. *Source:* Craig Riesterer, *Oakland Tribune.*

water-saturated. (See Figure 12-10.) Historic records indicate that such areas are vulnerable to heavy damage, and it was quickly realized that this pattern was borne out in the Loma Prieta earthquake. Specific areas in San Francisco, Oakland, Santa Cruz, and Watsonville that were located on such unconsolidated soils sites proved to be the pockets of heaviest damage.

Marina District, San Francisco

One of the pockets of heaviest damage areas experienced in this earthquake was located in the San Francisco Marina District. This area, primarily residential, is located in an area originally prepared for the siting of the 1915 Panama-Pacific International Exposition soon after the 1906 San Francisco earthquake. The only building remaining from this exposition is the Palace of Fine Arts designed by Bernard Maybeck. The rest of the buildings were demolished after the exposition closed, with the debris used as artificial fill along the historic bay margins to expand the area into the Marina District as it is known today, which includes the Marina Green, the Yacht Harbor, and blocks of residential developments.

Because of the poor quality of this artificial fill, as expected, severe building damage, liquefaction, fire, and earthquake-induced subsidence occurred throughout this area. Although buildings in the area were principally light wood-frame structures that have a good seismic performance record, building damage and consequential effects caused by the earthquake were dramatic. There is a tentative plan to create an earthquake memorial on one of the sites where houses were demolished.

FIRE FOLLOWING EARTHQUAKE

One of the most critical problems faced immediately in this district during the earthquake was the outbreak of a serious fire aggravated by the combustible light wood-frame construction. The fire was ignited by an explosion of natural gas in an apartment unit when gas lines ruptured. (See Figure 12-11.) It quickly turned into a four-alarm fire; some responding units were called in from Marin County, and fortunately they were able to use the Golden Gate Bridge, which was not damaged by the earthquake.

Through efforts that extended into the night and early morning, the fire was confined to one block, with residents of the area volunteering to help pull in water hoses from the San Francisco Bay waters. Again fortunately, in comparison to the 1906 San Francisco earthquake, there weren't any significant winds to fan and spread the flames. By early Wednesday morning the fire was under control and by 12:30 P.M. it had been completely extinguished with no further danger of flare-up. Accordingly, a repeat of the historic 1906 conflagration following the earthquake was avoided.

Figure 12-11 Fire in Marina District following the 1989 Loma Prieta earthquake. *Source:* Tom Duncan, *Oakland Tribune*.

BUILDING LAYOUT AND CONFIGURATION

In the Marina District, building layout and configuration played a significant role in the seismic performance of buildings, as seen by many of the damage patterns. The first-story collapse of several four-story wood-frame corner apartment/condominium units pierced by garage doors on the ground floor along two sides of the building represents excellent case studies of the "soft-story" effect. As shown in Figures 12-12 and 12-13, dramatic failures of the first-floor "soft story" of these four-story corner buildings caused them either to (1) collapse upon themselves in telescope fashion; (2) fall into the street; or (3) sag precariously over the sidewalk and street, with the "soft story" dangerously failing and distorted at a 45-degree angle.

In the Marina District, owing to the failure of poor soils, subsidence was clearly in evidence. Numerous sidewalks in the area buckled, gas pipelines ruptured, and water distribution lines were severely affected. (See Figure 12-14.) Many three-story row houses along Marina Boulevard and other streets in the district sunk straight down 3 to 6 in., making it impossible to open garage doors on the ground floor.

Another aspect of building damage in this area was related to the deterioration of wood sheathing under the exterior plaster, brick veneer, or wood siding; decay and moisture had been introduced

Figure 12-12 Telescoped four-story apartment building collapsing into street, Marina District, San Francisco. *Source:* Tom Duncan, *Oakland Tribune.*

Figure 12-13 Four-story apartment building telescoped into three stories over collapsed first-floor soft story, Marina District, San Francisco. Note horizontal sheathing at second floor.

Figure 12-14 Buckled sidewalks due to liquefaction, Marina District, San Francisco.

Figure 12-15 Horizontal sheating on collapsed apartment building, Marina District, San Francisco.

into the wood-frame cavities over the years. Many of the houses in the area are of pre-1934 construction, built before the use of plywood for shear walls was developed. Thus, their wood sheathing in general consists of 1 × 6 wood members laid horizontally, which is not very effective in resisting lateral loads. (See Figure 12-15.) As indicated earlier, the poor soils in the Marina District and inadequate building foundations didn't help building performance either.

Architectural Nonstructural Aspects

Once again, damage control relative to the failure of nonstructural building elements was an important architectural consideration. Some large corporations are reporting up to $45/50 million property damage, a considerable amount of which is directly attributable to nonstructural failures.

Several significant examples of nonstructural damage patterns are in evidence. A dramatic illustration of the relationships between nonstructural failure and life safety concerns is found in the ceiling failure of the Geary Theater in San Francisco. Only the basic structural supporting system remains of the proscenium arch ceiling; its nonstructural elements rest in a pile of rubble on the floor along

Figure 12-16 Interior of Geary Theater, San Francisco, October 1989 Loma Prieta earthquake. *Source:* Frederick Larson, *San Francisco Chronicle.*

with the fallen ceiling lighting grid and mounds of ceiling plaster, which cover the first six rows of seats in the auditorium. (See Figure 12-16.) Luckily, the theater, which can hold 1350 persons, was not in use at the time of the earthquake.

Building Contents

A personal computer software manufacturer in Scott's Valley, about 5 miles from the epicenter, incurred damage when the water piping and sprinkler systems ruptured in its main buildings. Water poured over equipment, assembly lines, and product components, forcing all employees out into the parking lot, where they continued their critical work. Fortunately, it was generally warm weather and not raining, so the employees were able to continue their production capacities outdoors for two full days before the mess was cleaned up inside the buildings.

Merchandise in several major department stores, supermarkets, retail wine outlets, and boutique shops throughout the area were typically overturned or thrown off shelves and racks onto the floors. One department store reported a $10 million dollar loss in reported income due to the earthquake (this total also included loss of sales due to building function impairment and the scarcity of customers following the event).

The Asian Art Museum in San Francisco lost more than $10 million worth of irreplaceable art pieces. The majority of damaged artifacts were from the Southeast Asian Khmer Empire collection from the sixth and eleventh centuries A.D. Several stone sculptures were thrown off their display pedestals and smashed on the floor. The most valuable work destroyed by the earthquake was a thirteenth century B.C. Shang dynasty ceremonial wine vessel, similar to a piece recently sold at a Sotheby's auction for $3 million. Shattered also beyond repair was a second century A.D. Hindu stone sculpture, Nagaraja, leaving only one other such piece in the world. Structurally, the museum building only experienced minor cracks in the concrete floor, and none of the glass display cases in the galleries was broken.

Performance of Hospitals

In California, the Hospital Act of 1972 required that major hospitals be designed to mandated performance standards to ensure their continued function and services and prevent crippling nonstructural damage during severe earthquakes. This earthquake was a good test of hospitals designed under the new 1972 act.

New hospitals performed well, with little or no interruption, since damage to the basic structural system, contents, and nonstructural components was minimal, especially in shear wall buildings. Emergency generators kicked in as planned to supply power, as needed, to keep the hospitals operational.

High levels of damage were evident in older pre-1972 hospitals still in use. A seven-story annex to a major hospital constructed in 1927 in Oakland was severely damaged and had to be evacuated. Critical damage to nonstructural components also rendered the hospital nonfunctional. Two stories of the Santa Clara Valley Medical Center were evacuated due to structural damage. Watsonville Community Hospital suffered moderate structural damage, and the fourth floor was evacuated owing to loss of elevators and exteriors windows. Damage to the Palo Alto Veterans Administration hospital facilities has been estimated at $30 million, and two of the six buildings on the site were evacuated indefinitely. The entire Stanford Medical Center suffered about $4 million in damage.

Performance of Public Schools

Schools designed under Field Act standards performed well structurally. There were no spectacular failures or collapses. Some temporary, portable classrooms were knocked off their supports, but they were quickly picked up, put back on their supports, and were again in use after two to three days. A preliminary survey of the

1544 public schools in the earthquake-impacted region was made to estimate damage. Only three schools in the entire region, in San Francisco, Watsonville, and Los Gatos, sustained severe damage.

The only San Francisco school to suffer severe structural damage was O'Connell High School, which officials say may cost the district as much as $10 million to repair. The district bought the building, originally a warehouse, in the 1950s and later converted it into a high school. Oakland's schools fared better, with about $1.5 million in damage.

The performance of nonstructural components was mixed. Light fixtures performed well and did not break loose from their ceiling mounts. Damage to bookcases and library stacks was extensive, indicating that additional attention is required in this area. Fortunately, at 5:04 P.M., the schools were not in session, so there weren't any critical injuries due to overturned bookcases or stacks.

Performance of Unreinforced Masonry (URM) Buildings

Older, existing pre-1934 unreinforced masonry buildings did not perform well during this earthquake, which had high acceleration characteristics. About 10 deaths were attributed to the collapse or partial collapse of buildings in this class. The worst damage to URM buildings occurred in Oakland, Los Gatos, Santa Cruz, Watsonville, and the South of Market District in San Francisco.

It is said that the downtown center of Santa Cruz, which had many URM buildings, will never be the same. Damage to the older URM buildings here was extensive. At least half of the older commercial buildings on Pacific and Front streets were condemned and scheduled for demolition. The Pacific Garden Mall, which contained many URM buildings, was turned into rubble (see Figure 12-17). Clouds of dust rose from several blocks in the downtown area after the earthquake struck. Four people were killed in the collapse of brick walls in this area. In San Francisco, six died when an unreinforced exterior brick wall of a four-story URM building collapsed onto their cars as they were leaving work.

Rehabilitation methods used in the seismic retrofit of URM buildings in the region were generally effective, but in several cases did not preclude critical damage. The 1989 Loma Prieta earthquake demonstrates that much more attention must be given to URM buildings located in areas of high seismic risk around the country. As many of these structures are historic buildings, their complete abatement by demolition is not the answer. Other solutions must be found to save them architecturally and structurally as strategic components of the urban fabric.

Homelessness Following the Earthquake

As indicated, about 13,000 people were rendered homeless by the earthquake, with some estimates reaching as high as 20,000. In

Figure 12-17 Damaged URM buildings in Pacific Garden Mall, downtown Santa Cruz, California. *Source:* Roy H. Williams, *Oakland Tribune.*

Watsonville alone, approximately 2000 residences were lost. San Francisco has determined that it will take roughly two years and cost about $191 million to replace and rebuild the estimated 5100 housing units that were lost. Over 1000 dwelling units were destroyed in Oakland. As a result of this earthquake, some building departments in the region are seeking legislation requiring all owners of one- and two-unit wood-frame residences to bolt frames of their houses to their foundations and brace cripple walls. In the Mission District of San Francisco, buildings that are located over the former Mission Creek continue to sink and settle into the landfill, and other housing units are expected to require demolition.

Public school buildings, county fairground facilities, and national guard armories, all of which performed quite well during the earthquake, served as emergency housing shelters. About 200 families from the Watsonville area were housed in exhibition halls on the country fair grounds. Many of the temporary shelters were operated by the Red Cross. The Marina Middle School successfully served as a shelter for the homeless and command post for the hard-hit Marina residential district in San Francisco.

Because of the shortage of housing units, Watsonville set up a tent city that operated for several weeks in a small park near the downtown area. Many of the homeless preferred this arrangement, which allowed them to be near their damaged homes, to the trailers offered to them, which were located on the outskirts of the city.

In the hardest hit area of the Santa Cruz Mountains, county supervisors temporarily banned new construction in an area near the Loma Prieta mountain summit after learning that the entire region was vulnerable to massive landslides. Later it was reported that water wells were sheared off at 65- to 160-ft depths, although initial appearances of the area had only indicated surface cracking in extensive circular patterns that looked like simple landslides. This indicates that something might be happening deep in the mountains. About 600 building sites are in the slide area affected by the building ban.

Economic Impact

Estimates of total property loss attributed to the Loma Prieta earthquake have reached a range as high as $8 to $10 billion. Congress passed legislation to provide over $3.4 million in federal aid to support earthquake relief efforts. The state of California passed emergency legislation to increase sales taxes by an initial quarter of a cent for 13 months to provide about $800 million for earthquake recovery in the San Francisco Bay region.

Whereas the total number of collapses is low, the total damage estimate is quite high. For example, while there was no spectacular destruction at the Stanford University campus in Palo Alto, damage is judged to be as high as $160 million. Total losses in the Santa Cruz area are estimated to be about $170 million.

Although final dollar loss estimates are yet to be derived, it is clear from actions already taken that the costs of this earthquake are substantial. The region hit by the earthquake, a seven-county area, will take a significant length of time to recover from this economic loss.

SUMMARY

The damage patterns of the 7.1 magnitude Loma Prieta earthquake, as compared to other seismic events that resulted in total or near-total destruction (e.g., Armenia, Tangshan, Managua, and Skopje), present significant opportunities for participation by architects and planners in rebuilding efforts and in-depth studies of architectural issues. For example, in the Marina District, where eight buildings have been demolished and another 249 posted unsafe by the Department of Public Works, one prominent architect has already proposed a tentative scheme to the Planning Department for turning the Marina into a waterfront park and relocating residents to the Presidio of San Francisco.

In the Marina, however, despite the proposal of bold relocation plans, thousands of residents are still struggling with the disaster and attempting to rebuild their lives. It is believed by many planners that wholesale clearance of the Marina District isn't even a choice open for consideration. Although there may be some empty lots in the district, past experiences in postearthquake reconstruction efforts indicate that comprehensive proposals to rebuild the Marina in its present location will be considered and adopted. In all probability, master plans will be introduced and accepted to preserve the Marina's architectural integrity in the construction of new buildings where required by recovery goals. Already an ordinance has been adopted that permits owners in the Marina to rebuild to the same size as before the earthquake.

More than ever before, the Loma Prieta earthquake again reinforces the inevitable conclusion that architects and planners must be prepared to have a major hand in postearthquake recovery efforts. The housing problem alone, resulting from the loss of over 9500 housing units throughout the region, requires participation by architects and planners. The challenge is to rebuild affordable dwellings, which were already in acute shortage before the earthquake.

Since the 1983 Coalinga earthquake in California, when the California Council of the American Institute of Architects (CCAIA) sent in an architectural reconnaissance team, the profession has become active in postearthquake recovery efforts. The CCAIA sent another reconnaissance team to investigate Mexico City's devastating 1985 earthquake. And in 1988, the CCAIA combined forces with the national AIA to send an advisory team to Armenia to help in the development of plans for the reconstruction of Spitak and Leninakan. With the lead taken by the AIA, it is clear that more architects must assume responsibilities associated with earthquake hazards mitigation efforts and rebuilding plans. More than ever, participation of architects and planners in achieving seismic safety goals is required to meet the complexities of today's society.

REFERENCES

Algermissen, S. T. (1989). *Earthquake Parameters and Effects: Relationships to Earthquake Casualties.* U. S. Geological Survey, Denver, Colorado.

Algermissen, S. T. Rinehart, W. A., Dewey, J., Degenkolb, H. J., Cluff, L., McClure, F. E., Scott, S., Gordon, R. F., Lagorio, H. J., Steinbrugge, K. V. (1972). *"A Study of Earthquake Losses in the San Francisco Bay Area: Data and Analysis",* A report prepared for the Office of Emergency Preparedness. National Oceanic and Atmospheric Administration, Environmental Research Laboratories, U.S. Department of Commerce, Washington, D.C.

Arnold, C. (1989). *"Mission to Armenia."* In *Architecture,* pp. 99–105. American Institute of Architects (AIA), Washington, D.C.

Arnold, C., and Lagorio, H. J. (1987). *"Chinese City Starts Over After Quake."* In *Architecture,* pp. 83–85. American Institute of Architecture (AIA), Washington, D.C.

Arnold, C., and Reitherman, R. (1982). *Building Configuration and Seismic Design.* Wiley, New York.

Astaneh, A., Bertero, V. V., Bolt, B. et al. (1989). *Preliminary Report on the Seismological and Engineering Aspects of the October 17, 1989, Santa Cruz (Loma Prieta) Earthquake,* Report No. UCB/EERC-89/14. Earthquake Engineering Research Center (EERC), University of California, Berkeley.

Ayres, J. M., Sun, T., and Brown, F. (1973). *"Nonstructural Damage to Buildings."* In *The Great Alaska Earthquake of 1964,* pp. 346–456. National Academy of Sciences (NAS) and National Research Council (NRC). Washington, D.C.

Blair, M. L. (1979). *Seismic Safety and Land-Use Planning: Selected Examples from California. Geological Survey Professional Paper (U.S.)* **941-B.**

Bolt, B. A., Horn, W. L. et al. (1977). *Geological Hazards,* 2nd rev. ed. Springer-Verlag, New York.

Borcherdt, R. D., Ed. (1975). *"Studies for Seismic Zonation of the San Francisco Bay Region". Geological Survey Professional Paper (U.S.)* **941-A.**

Borcherdt, Roger D., Donavan, N. C. et al. (1989). *Geoscience Investigations of the Earthquake of October 17, 1989, Near the Summit of Loma*

Prieta in the Southern Santa Cruz Mountains, Special Report. EERI Geoscience Reconnaissance Team, Earthquake Engineering Research Institute (EERI), El Cerrito, CA.

Botsai, E. E., Eberhard, J. P., Lagorio, H. J. et al. (1976). *Architects and Earthquakes.* National Science Foundation and the AIA Research Corporation, Washington, D.C.

Building Seismic Safety Council (BSSC) (1988). *NEHRP Recommended Provisions for the Development of Seismic Regulations for New Buildings.* Applied Technology Council (ATC), Redwood City, CA, and the Federal Emergency Management Agency (FEMA), Washington, D.C.

Casada, F. N. (1987). *Manual de Diseno y Construccion de Viviendas Para Personas de Escasos Recursos.* Secretaria General de Obras, Departamento del Distretto Federal, Mexico City, Mexico.

Chopra, A. K., (1982). *Dynamics of Structures—A Primer.* Earthquake Engineering Research Institute (EERI), Berkeley, CA.

Comerio, M., Friedman, H., and Lagorio, H. J. (1987). *Unreinforced Masonry Seismic Strengthening Workshop and Cost Analysis.* Center for Environmental Design Research (CEDR), University of California, Berkeley, and Center for Environmental Change (CEC), Inc., San Francisco.

Conner, H. W., Harris, J. R. et al. (1987). *Guide to Application of the NEHRP Recommended Provisions in the Earthquake-Resistant Building Design.* Federal Emergency Management Agency (FEMA) and Building Seismic Safety Council (BSSC), Washington, D.C.

Cornell, J. C. (1978). *The Great International Disaster Book.* Pocket Books, New York.

Crawley, S. W., et al. (1987). *The Architect's Study Guide to Seismic and Lateral Loads in Architectural Design.* American Institute of Architects (AIA), Washington, D.C.

Department of Defense Tri-Services Seismic Design Committee (1982). *Technical Manual: Seismic Design for Buildings.* Departments of the Army, Navy, and the Air Force, Office of the Chief Engineers, Washington, D.C., for U.S. Army Division Engineer, South San Francisco, CA.

Dynes, R. R., *Reconstruction in the Context of Recovery: Thoughts on the Alaskan Earthquake.* Department of Sociology, University of Delaware, Newark.

Eisner, R. (1988). *Land Use and Planning Lessons of the 1985 Mexico City Earthquake,* pp. 209–326. Architectural and Urban Design Lessons from the 1985 Mexico City Earthquake, Colegio de Arquitectos de Mexico and the Council on Architectural Research, American Institute of Architects (AIA) and the Association of Collegiate Schools of Architecture (ACSA), Washington, D.C.

Eisner, R. (1989a). *"The Status of Earthquake Prediction."* In *NETWORKS: Earthquake Preparedness News,* Vol. 4, No. 1. Bay Area Regional Earthquake Preparedness Project (BAREPP), Oakland, CA.

Eisner, R. (1989b). *Earthquake Vulnerability Analysis for Local Governments,* Special Report. Bay Area Regional Earthquake Preparedness Project (BAREPP), Oakland, CA.

Elsesser, E. (1984). "Life Hazards Created by Nonstructural Elements." In *Nonstructural Issues of Seismic Design and Construction,* Workshop Proceedings pp. 27–36. Earthquake Engineering Research Institute (EERI), Berkeley, CA.

Escalante, L. (1986). "Performance of Lifelines", pp. 429–448, the 1985 Chile Earthquake, In *Earthquake Spectra,* Vol. 2. No. 2, Earthquake Engineering Research Institute (EERI), El Cerrito, California.

Federal Emergency Management Agency (FEMA) (1980). Committee on Assessment of Consequences and Preparation of Consequences and Preparations for a Major California Earthquake, *An Assessment of the Consequences and Preparations for a Catastrophic California Earthquake: Fingings and Actions Taken.* FEMA, Washington, D.C.

Geis, D. E., and Smith, K. N. (1989a). *Architectural and Urban Design Lessons from the 1985 Mexico City Earthquake,* AIA, Association of American Schools of Architecture (AIA/ACSA), Washington, D.C., and Colegio de Arquitectos de Mexico/Sociedad de Arquitectos Mexicanos, Mexico City.

Geis, D. E., and Smith, K. N. (1989b). *Designing for Earthquakes in Southern California.* AIA, Association of Collegiate Schools of Architecture (AIA/ACSA), Washington, D.C.

Greene, M. R. (1987). "Skopje, Yoguslavia: Seismic Concerns and Land Use Issues During the First Twenty Years of Reconstruction Following a Devastating Earthquake." In *Earthquake Spectra,* Vol. 1, No. 2, pp. 103–118. Earthquake Engineering Research Institute (EERI), El Cerrito, CA.

Griggs, G. B., and Gilchrist, J. A. (1983). *Geological Hazards, Resources, and Environmental Planning,* 2nd ed. Wadsworth, Belmont, CA.

Guevara, L. T. (1989). "Architectural Considerations in the Design of Earthquake-Resistant Buildings: Influence of Floor-Plan Shape on the Response of Medium-Rise Housing to Earthquakes." Ph.D. Dissertation, Graduate Division, University of California, Berkeley.

HABITAT (United Nations Centre for Human Settlements), (1988). *Seismic Risk Management in the Planning of the Historic Center of Mexico City,* Project Monograph. HABITAT, Nairobi, Kenya.

Handler, P. (1973). *The Great Alaska Earthquake of 1964: Engineering.* National Academy of Sciences (NAS), Washington, D.C.

Hauf, H. (1973). "Architectural Factors in Earthquake Resistance." In *The Great Alaska Earthquake of 1964,* pp. 340–345. National Academy of Sciences (NAS) and National Research Council (NRC). Washington, D.C.

Helfant, D. B. (1989). *Earthquake Safe: A Hazard Reduction Manual for Homes.* Builders Booksource Publication, Berkeley, CA.

Iglesias, J. (1989). *Reconstruction of Mexico City After the 1985 Earthquake.* Department of Materials and Construction, Universidad Autonoma Metropolitana (UNAM), Mexico City, Mexico.

International Conference of Building Officials (1988). *Uniform Building Code.* Whittler, CA.

Jones, B. G. (1989). *The Need for a Dynamic Approach to Planning for Reconstruction After Earthquakes.* Department of Planning, Cornell University, Ithaca, NY.

Kessler, J. J. (1973a). *Hospitals and Medical Facilities,* Vol. 1, Part A, pp. 175–176. National Oceanic and Atmospheric Administration (NOAA), Department of Commerce, Washington, D.C. (1983).

Kessler, J. J. (1973b). *Summary and Conclusions for Hospitals and Medical Facilities,* Vol. 1, Part A, pp. 295–296. The San Fernando, California, Earthquake of February 9, 1971 National Oceanic and Atmospheric Administration (NOAA), Department of Commerce, Washington, D.C.

Ketter, (1987). *1986–1987 Annual Report.* National Center for Earthquake Engineering Research (NCEER), Buffalo, NY.

Klain, M., Ricci, E., Safar, P., Comfort, L. et al. (1989). "Disaster Reanimatology Potentials: A Structured Interview Study in Armenia: Methodology and Preliminary Results." *Prehospital and Disaster Medicine* 4 (2), 135–154.

Lagorio, H. J. (1985). "Seismic Performance of Critical, Emergency Service Facilities." In *Proceedings of the Joint U.S.–Romanian Seminar on Earthquakes and Energy,* pp. 323–339. INCERC, Building Research Institute, Bucharest.

Lagorio, H. J. (1986). *Improving Seismic Safety: Urban Reconstruction Planning Following Earthquakes.* Center for Environmental Design Research (CEDR), University of California, Berkeley.

Lagorio, H. J., and Botsai, E. E. (1978). "Urban Design and Earthquakes." In *Proceedings of the Second International Conference on Microzonation for Safer Construction: Research and Application,"* Vol. 1, pp. 193–202. University of Washington, Seattle.

Lagorio, H. J., and Gavarini, C. (1984). *Joint USA/Italy Workshop on Seismic Repair and Retrofit of Existing Buildings.* Institute of Building Sciences, University of Rome, and Center for Environmental Design Research (CEDR), University of California, Berkeley.

Lagorio, H. J., Friedman, H., and Wong, K. (1986). *Issues for the Seismic Strengthening of Existing Buildings: A Practical Guide for Architects.* Center for Environmental Design Research (CEDR), University of California, Berkeley.

Luft, R. W. (1989). "Comparisons Among Earthquake Codes." In *Earthquake Spectra,* Vol. 5, No. 4, pp. 767–789. Earthquake Engineering Research Institute (EERI), El Cerrito, CA.

Mader, G. G. (1980). *Land Use Planning After Earthquakes.* William Spangle and Associates, Inc., Portola Valley, CA.

Mader, G. G., and Lagorio, H. J. (1981). *Earthquake in Campania-Basilicata, Italy, November 23, 1980: Architectural and Planning Aspects.* Earthquake Engineering Research Institute (EERI), Berkeley, CA.

Mahin, S. (1987). *Observations from Recent Earthquakes Regarding the Design of Buildings,* Technical Paper. Continuing Education in Engineering, Summer Seminar Series, University of California Extension Division, Berkeley.

Mallet, R. (1862). *Great Neapolitan Earthquake of 1857,* 2 vols, London.

McCue, G. M. (1976). *The Interaction of Building Components During Earthquakes.* McCue Boone Tomsick (MBT), San Francisco, and Engineering Decision Analysis Co. Inc., Palo Alto, CA.

Meehan, J. F., Cluff, L. S. et al. (1973). *Managua, Nicaragua, Earthquake of*

December 23, 1972. Earthquake Engineering Research Institute (EERI), Berkeley, CA.

Murphey, L. M. (1973). Scientific Coordinator, *The San Fernando, California, Earthquake of February 9, 1971,* Vols. I–III. National Oceanic and Atmospheric Administration (NOAA), Department of Commerce, Washington, D.C.

Noji, E. (1989). *"Medical and Healthcare Aspects,"* Armenia Earthquake Reconnaissance Report. In *Earthquake Spectra, Special Supplement,* pp. 161–169. Earthquake Engineering Research Institute (EERI), El Cerrito, CA.

Olson, R. A., et al. (1987). *Data Processing Facilities: Guidelines for Earthquake Hazards Mitigations.* VSP Associates, Inc., and FIMS, Inc., Sacramento, Ca.

Ramos, H. P. (1987). *"El Nuevo Color del Centro de la Ciudad de Mexico." Mexico Desconocida* No. 126.

Reitherman, R. K. (1983). *Reducing the Risks on Nonstructural Earthquake Damage: A Practical Guide.* Scientific Services, Inc., Redwood City, CA.

Reitherman, R. K. (1989). *"Significant Revisions in Model Seismic Code."* In *Architecture,* pp. 106–112. American Institute of Architecture (AIA), Washington, D.C.

Rojahn, C., and Reitherman, R. K. (1989). *Procedures for Postearthquake Seismic Evaluation of Building.* ATC-20, Applied Technology Council (ATC), Redwood City, CA.

Schiff, A. J. (1988). "Response of Lifelines and their Effect on Emergency Response," pp. 339–366, The 1987 Whittier-Narrows Earthquake, *Earthquake Spectra,* Vol. 4, No. 2, Earthquake Engineering Research Institute (EERI), El Cerrito, CA.

Scholl, R., Ed. (1982). *Fix'Em: Identification and Correction of Deficiencies in Earthquake Resistance of Existing Buildings.* Earthquake Engineering Research Institute (EERI), Berkeley, CA.

Scholl, R., Ed. (1986). *Reducing Earthquake Hazards: Lessons Learned From Earthquakes.* Earthquake Engineering Research Institute (EERI), El Cerrito, CA.

Scholl, R. and Lagorio H. J., Eds. (1984). *Nonstructural Issues of Seismic Design and Construction,* Workship Proceedings. Earthquake Engineering Research Institute (EERI), Berkeley, CA.

Scholl, R., and Stratta, J. L., Eds. (1984). *Coalinga, California, Earthquake of May 2, 1983.* Earthquake Engineering Research Institute (EERI), Berekeley, CA.

Schwartz, E. (1981). *"City of Los Angeles Ordinance for the Abatement of Existing Hazardous Buildings."* In *Proceedings of PRC/USA Joint Workshop on Earthquake Disaster Mitigation.* Ministry of Urban and Rural Construction and Environmental Protection (MURCEP), Beijing, China, and Center for Environmental Design Research (CEDR), University of California, Berkeley.

Scott, S. (1985). *Rehabilitating Hazardous Masonry Buildings: A Draft Model Ordinance.* California Seismic Safety Commission (CSSC), Sacramento, CA.

Sharpe, R., and Culver C. (1978). *Tentative Provisions for the Development of Seismic Regulations for Buildings.* Applied Technology Council (ATC), Palo Alto, CA, and the National Bureau of Standards (NBS), Washington, D.C.

Shea, G. H., Ed. (1988). *Unreinforced Masonry Buildings: The Threat to Every Community,* Special Report. Bay Area Regional Earthquake Preparedness Project (BAREPP), Oakland, CA.

Seed, R. et al. (1990). *Preliminary Report on the Principal Geotechnical Aspects of the October 17, 1989 Loma Prieta Earthquake.* Earthquake Engineering Research Center, Berkeley, CA.

Steinbrugge, K. V. (1982). *Earthquakes, Volcanoes, and Tsunamis: An Anatomy of Hazards.* Skandia America Group, New York.

Steinbrugge, K. V., and Lagorio, H. J. (1985). *Relative Earthquake Safety in Buildings,* pp. 137–139. California Geology, Division of Mines and Geology (DMG), Department of Conservation, Sacramento, CA.

Steinbrugge, K. V. et al. (1979). *Existing Hazardous Buildings: Assessing Direct Post-Earthquake Impacts,* CSSC 79-01. California Seismic Safety Commission (CSSC), Sacramento.

Stratta, J. L. (1984). *"The Concern With Nonstructural Damage."* In *Nonstructural Issues of Seismic Design and Construction,* Workshop Proceedings, pp. 3–23. Earthquake Engineering Research Institute (EERI), Berkeley, CA.

Stratta, J., L. (1987). *Manual of Seismic Design.* Prentice-Hall, Englewood Cliffs, NJ.

Thiel, C. C., Ed. (1988a). "The 1985 Mexico Earthquake." In *Earthquake Spectra,* Part A, Vol. 4, No. 3. Earthquake Engineering Research Institute (EERI), El Cerrito, CA.

Thiel, C. C., Ed. (1988b). "The 1985 Mexico Earthquake." In *Earthquake Spectra,* Part B, Vol, 4, No. 4. Earthquake Engineering Research Institute (EERI), El Cerrito, CA.

Thiel, C. C., Ed. (1989). "The 1985 Mexico Earthquake." In *Earthquake Spectra,* Part C, Vol. 5, No. 1. Earthquake Engineering Research Institute (EERI), El Cerrito, CA.

Tobriner, S. (1982). *The Genesis of Noto: An Eighteenth-Century Sicilian City.* University of California Press, Berkeley.

Vitaliano, D. B. (1973). *Legends of the Earth: Their Geologic Origins.* Indiana University Press, Bloomington.

Wallace, R. E., Davies and McNally, K. (1984). United States Earthquake Prediction Terminology, Geological Survey, Menlo Park, CA.

Wang, M. (1987). *"Cladding Performance on a Full Scale Test Frame."* In *Earthquake Spectra,* Vol. 3, No. 1, pp. 119–173. Earthquake Engineering Research Institute (EERI), El Cerrito, CA.

Wong, K. M. (1987). *"Seismic Strengthening of Unreinforced Masonry Buildings".* Center for Environmental Design Research (CEDR), University of California, Berkeley.

Woodhouse, J. (1989), Article on Imaging the Earth, *San Francisco Chronicle,* Sunday Science Section.

INDEX

POINT LOMA NAZARENE COLLEGE
RYAN LIBRARY